Studies in Applied Philosophy, Epistemology and Rational Ethics 4

Editor-in-Chief

Prof. Dr. Lorenzo Magnani
Department of Philosophy
University of Pavia
Piazza Botta 6
27100 Pavia
Italy
E-mail: lmagnani@unipv.it

Editorial Board

Prof. Atocha Aliseda
Instituto de Investigaciones Filosoficas, Universidad Nacional Autónoma de México (UNAM), Ciudad Universitaria, Coyoacan 04510, Mexico D.F.
E-mail: atocha@filosoficas.unam.mx

Prof. Giuseppe Longo
Laboratoire et Departement d'Informatique, CREA, CNRS and Ecole Polytechnique, LIENS 45, Rue D'Ulm, 75005 Paris, France
E-mail: Giuseppe.Longo@ens.fr

Prof. Chris Sinha
Department of Psychology, University of Portsmouth, King Henry Building, King Henry I Street, Portsmouth, Hampshire PO1 2DY, UK
E-mail: chris.sinha@port.ac.uk

Prof. Paul Thagard
Department of Philosophy, Faculty of Arts, Waterloo University, Waterloo, ON N2L 3G1, Canada
E-mail: pthagard@watarts.uwaterloo.ca

Prof. John Woods
Department of Philosophy, University of British Columbia, 1866 Main Mall BUCH E370, Vancouver, BC V6T 1Z1, Canada
E-mail: jhwoods@interchange.ubc.ca
E-mail: woods@uleth.ca

For further volumes:
http://www.springer.com/series/10087

Massimo Negrotti

The Reality of the Artificial

Nature, Technology and Naturoids

Massimo Negrotti
DISC-LCA, Lab for the Culture
of the Artificial
Università di Urbino 'Carlo Bo'
Urbino
Italy

ISSN 2192-6255 ISSN 2192-6263 (electronic)
ISBN 978-3-642-44353-4 ISBN 978-3-642-29679-6 (eBook)
DOI 10.1007/978-3-642-29679-6
Springer Heidelberg New York Dordrecht London

© Springer-Verlag Berlin Heidelberg 2012
Softcover reprint of the hardcover 1st edition 2012
This work is subject to copyright. All rights are reserved by the Publisher, whether the whole or part of the material is concerned, specifically the rights of translation, reprinting, reuse of illustrations, recitation, broadcasting, reproduction on microfilms or in any other physical way, and transmission or information storage and retrieval, electronic adaptation, computer software, or by similar or dissimilar methodology now known or hereafter developed. Exempted from this legal reservation are brief excerpts in connection with reviews or scholarly analysis or material supplied specifically for the purpose of being entered and executed on a computer system, for exclusive use by the purchaser of the work. Duplication of this publication or parts thereof is permitted only under the provisions of the Copyright Law of the Publisher's location, in its current version, and permission for use must always be obtained from Springer. Permissions for use may be obtained through RightsLink at the Copyright Clearance Center. Violations are liable to prosecution under the respective Copyright Law.
The use of general descriptive names, registered names, trademarks, service marks, etc. in this publication does not imply, even in the absence of a specific statement, that such names are exempt from the relevant protective laws and regulations and therefore free for general use.
While the advice and information in this book are believed to be true and accurate at the date of publication, neither the authors nor the editors nor the publisher can accept any legal responsibility for any errors or omissions that may be made. The publisher makes no warranty, express or implied, with respect to the material contained herein.

Printed on acid-free paper

Springer is part of Springer Science+Business Media (www.springer.com)

"Ad Alessandro e Angelo Maria e ai loro sguardi sorridenti"

Contents

Part IV

Part V

Part I

Chapter 1
Daedalus and Icarus

History and mythology agree to assign to the human ambition to build objects inspired by nature a very ancient origin. It is certain that the ambition to reproduce and improve natural objects and events constitutes for humans a sort of constant goal, almost an imperative whose achievement seems to be linked not only to practical utility, but also to people's deepest psychology. Furthermore, when the reproduction takes the form of artistic, literary, or musical representation, the representational bias clearly shows how the externalization of one's visions of the world is definitely a universal need for human beings.

Nevertheless, the reproduction of something natural is never an easy enterprise, and, as we shall see analytically in this work, the actual replication of a natural object or process by means of some technology is quite impossible. Therefore, it seems to be reasonable to introduce the term *naturoid*[1] to designate any real artifact coming from the attempt to reproduce natural instances.

One of the most ancient naturoids, though only imagined, is surely that of the artificial wings designed for Icarus by his father, Daedalus, who, according to classic mythology, came crashing down near Samo, because the wax that fixed his wings had melted. Icarus probably repented bitterly for not having followed the advice of Daedalus, who had warned him not to get too close to the Sun. While a bird, in a similar circumstance, might only have been singed and would soon have descended to more reasonable heights, the artificial wings of Icarus could not withstand the test, and he died suddenly.

On the other hand, we know that Homer, in the *Iliad*, describes the god, Efesto, who creates the first woman, Pandora, from clay; Plato speaks of the mobile statue built by Daedalus himself; Argonauts, who searched for the Golden Fleece, had an artificial watchdog at their disposal.

---

[1] The author introduced the term 'naturoid' in the book 'Naturoids: On the Nature of the Artificial,' published in 2002 by World Scientific Publishing. Some material from that book is reproduced here with kind permission.

M. Negrotti, *The Reality of the Artificial*, Studies in Applied Philosophy, Epistemology and Rational Ethics 4, DOI: 10.1007/978-3-642-29679-6_1,
© Springer-Verlag Berlin Heidelberg 2012

Attempts to reproduce natural instances date back to antiquity. As has been noted (Boas 1927; Vogel 2000), such attempts are even found in the early forms of art and in several fantastic poetic or religious accounts. As stated by de Solla Price,

> Our story, then, begins with the deep-rooted urge of man to simulate the world about him through the graphic and plastic arts. The almost magical, naturalistic rock paintings of prehistoric caves, the ancient grotesque figurines and other 'idols' found in burials, testify to the ancient origin of this urge in primitive religion (de Solla Price 1964)

Furthermore, automata—reproductions of human beings very different from one another in terms of their substance—have abounded throughout the history of human imagination, starting with the Bible and including Faust and the *Rossum's Universal Robot* (R.U.R.) by Čapek.

It would surely be very easy to expose, case by case, the weak points of each of the 'machines' quoted above, but there is another much more compelling question. What kind of relationship exists between these attempts to imitate nature and the technology of that particular historical period? In other words, is technology intrinsically intended to reproduce something existing in nature, or is it developed with other aims in mind as well? After all, it is like when we ask ourselves: has man created technology only to reproduce nature?

The history of technology clearly shows that man, in designing and building objects, machines, and processes is often motivated by a desire to imitate, but in many other cases he aims to control and dominate natural events, rather than reproduce them by means of expedients which are refined to varying degrees. As the mathematician Henry de Monantheuil stated in the sixteenth century,

> … man, being God's image, was invited to imitate him as a mechanician and to produce objects which could cope with those made by Nature (de Monantheuil 1517, in Bredekamp 1993)

In order to control natural events one needs to know them and our technologies will vary in their effectiveness according to the accuracy of this knowledge. However, these technologies will not necessarily be designed to imitate the phenomena in question, but to adapt to nature in order to exploit its features and reach some useful goal.

Thus, knowledge of some physical laws allowed us to build machines, such as electrical or internal combustion engines, which enhance our ability to move physically. The advent of writing led to the invention of various writing tools which, today, are highly advanced thanks to computer technology. The knowledge and the exploitation of other natural laws made it possible to conceive, and then construct, systems like the cathode-ray tube which make the display of graphs or images, possible.

In all these examples, and in the many more one could easily cite, there is no attempt to imitate, but, rather, an attempt to invent, to make behaviors, effects, and events possible which, on the basis of our pure and simple natural condition, would not be attemptable.

To sum up, close to Icarus—and to all his descendants who constitute the world of the *artificialists*—the figure of Prometheus stands out. In giving fire to man, he awakened man's ability to invent, i.e., his ability to establish construction targets for objects or processes, as it were, additional and therefore different from those already existing in nature.

As we know, in doing so, unfortunately man created dangers and disasters. Prometheus himself was the first to pay, with tremendous torture, for having taken possession of fire, which was a prerogative of the gods. Nevertheless, imitation and invention are two distinct circumstances and actions, and they require analyses which are likewise different.

In the following pages, we shall investigate and outline some fundamental aspects of the first of the above technologies, namely the *technology of naturoids*. A clear distinction is drawn between the technology of naturoids and *conventional technology*. Unlike the latter, the former implicitly or explicitly aims to reproduce something existing in nature. This distinction may be applied to all technological traditions, from the most prescientific to the most advanced ones, including nanotechnology, rightly defined by Crandall and Lewis as

> a descriptive term for a particular state of our species' control of materiality (Crandall and Lewis 1997)

Nanotechnology, in fact, can be oriented either to reproduce natural things or processes, exhibiting different features, or to produce new objects or materials. Indeed, nanotechnology may be oriented to design nanomotors or even artificial white cells, or to emulate other biological nanostructures able to perform useful tasks within the human body, though

> For the manufacture of more sophisticated materials and devices, including complex objects produced in large quantities, it is unlikely that simple self-assembly processes will yield the desired results. The reason is that the probability of an error occurring at some point in the process will increase with the complexity of the system and the number of parts that must interoperate. (National Materials Advisory Board 2006)

The dichotomy between conventional technology and the technology of naturoids tracks a distinction that, though accepted as a fact on a commonsense basis, has never been clearly drawn, but it is very useful in trying to understand rationally a wide range of phenomena which are not just technological.

In fact, some authors seem to have touched the subject, calling by a special name—namely 'alternate realizations'—the machines designed to replace natural structures (Rosen 1993) or maintaining the opportunity, for the human species, to adopt, in manufacturing, a nature-oriented *mimicry* strategy (Benyus 1997). Even Herbert Simon, in his seminal work *"The Sciences of the Artificial"* (Simon 1969), gives no particular relevance to the term 'artificial,' limiting himself to recognize that, sometimes, artificial things coincide with 'synthesized' objects, i.e., with objects built by drawing inspiration from natural ones.

The need for a new name—namely, naturoids—comes from the fact that, traditionally, all technological devices are defined as artificial, quite neglecting the important teleological difference sketched above. Even during the long debate on the feasibility of artificial intelligence projects, the problem was in no way placed in a larger frame, that is to say that of the general methodological rules that characterize the human attempt to reproduce natural instances. Such rules have to be understood as a sort of 'action plane' that, although not explicit, intrinsically characterizes the ideation, the design and then the realization of any naturoid.

Of course, at some detailed level, such as the nanotechnological or the atomic, one can think that the distinction between natural and artificial things disappears. Actually, the processes that are going on at these levels, be they natural or man-made, tend often to be indistinguishable. Nevertheless, at a more human-scale level, things are different since humans have always tried to imitate, or to concretely reproduce natural objects *as they see them*—both through their eyes and indirectly through some observation tool—and sometimes even as they *imagine*. In this frame, though the distinction between natural and artificial is certainly a heuristic one, as it happens for any scientific criterion, it may be useful for understanding the design of a class of artificial things, namely naturoids, and not of the whole range of man-made artifacts. Probably, like quantum physics is needed to understand what happens in microworlds, whereas classical physics works well in macroworlds, a theory of naturoids could be useful at its proper level, whereas it may be inapplicable at other levels.

We may consider imitation and invention—i.e., the basic human qualities which generate the technology of naturoids and conventional technology—as human aptitudes which, on a social and cultural level, give rise to very different classes of behaviors and activities. Imitation, for instance, exhibits a variety of expressions which range from the socialization of children to fashion, and the spreading of cultural models (scientific, technological, ethical, religious, juridical, etc.).

Imitation may sometimes include a true reproduction activity that concerns, as a target to be imitated, even man-made things or events.

This has happened many times in the field of works of art, and this justifies the use of both the terms 'imitation' (when one tries to only mimic some aspects of the appearance of an object or a process) and 'reproduction' (when one tries to reproduce something adopting a more or less detailed analytical model of it). As applied to works of art, the term reproduction was discussed, perhaps for the first time, in an influential 1936 article by W. Benjamin, entitled "The Work of Art in the Age of Mechanical Reproduction" (Benjamin 1936). Benjamin's speaking of *reproduction of a man-made work*, allows us to see that a theory of naturoids may be useful even when the object to be reproduced is not natural. For example, in the field of acoustics, digital surround-sound effects may try to emulate the man-made effects of Gothic churches, just as stereo effects try objectively to reproduce the practice introduced by Giovanni Gabrieli, in the church of San Marco in Venice during the sixteenth century. Digital technology allowed the Japanese company Yamaha in the late 1970s to emulate an acoustic piano, although, for various intrinsic reasons, such a piano cannot be compared to, say, a real Steinway D and its 'aura,' to use Benjamin's word.

To sum up, imitation is a larger area than that of naturoids, because it includes all attempts to remake something already made by man, both for counterfeit aims or for improving an existing device or process. However, in both these cases—which are beyond the scope of this book—we presume that the 'action plane' of the designer remains the same.

Invention, on the other hand, manifests itself in innovative social behaviors—which, when they succeed, will be imitated, as the French sociologist Gabriel Tarde explained a century ago (Tarde 1890)—and also in various typologies of economic enterprises, in exploration activities and in the generation of new ideas.

In the technological domain, invention and conventional technology work assuming that humans can imagine and realize objects, processes or machines by adopting natural resources in order to achieve additional aims as compared to the basic natural ones, often conceived even as ways to act against nature (Bensaude-Vincent and Newman 2007). In fact, invention is a kind of activity often based on an abstract rational way of studying and controlling the world by grasping and exploiting its uniformities through means of mathematical models. In this regard, we should not forget that, as Poincaré rightly states,

> The genesis of the mathematical creation is … the activity in which the human mind seems to take very little from the external world, and in which it acts or seems to act only by itself and on itself … (Poincaré 1952)

and, therefore, conventional technology places itself in a domain where human imagination and natural laws meet, giving birth to devices which are not experienced within nature, though they should match its features.

Conversely, in our approach to the field of artificial things, that is to say to the naturoids domain, we shall concentrate on those particular fields of activities which place at their center, on the basis of a more general imitation 'instinct,' the reproduction of something existing in nature, and whose reproduction—through construction strategies which differ from the natural ones—man considers to be useful, appealing or in any case interesting.

While awaiting the new technologies which, according to Drexler's vision of nanotechnology, will allow us to

> build almost anything that the laws of nature allow to exist, (Drexler 1986)

we shall take into consideration the efforts of men who try to reproduce natural instances through 'macrotechnology' strategies, on the basis of analytical models they build for such instances.

Chapter 2
Artificiality and Naturoids

The use of the term 'naturoid' calls for the resolution of an ambiguity that involves the concept of 'artificial' in many contexts. From a linguistic standpoint, the term artificial (*artificiale* in Italian, *künstlich* in German, *artificiel* in French) covers a heterogeneous area which should be clarified before we proceed. In all languages, this concept seems to generically indicate all that is "man-made" or "not natural" and, at the same time, though more rarely, something which tries to imitate things existing in nature. Nevertheless, it is a fact that, while no one would speak of an 'artificial telephone', everyone understands the meaning of an 'artificial flower' quite well. We believe that this situation can be interpreted quite easily. Though it has never been rationally defined, the concept of artificial, as an adjective, refers to an object, process or machine which aims to reproduce some natural object or process. Since flowers exist in nature but not telephones, the adjective 'artificial' has no meaning if we attribute it to any object invented and built by man, i.e., an 'artifact', while it takes on full meaning when it is finalized to reproduce a natural object.

The Italian linguists Devoto and Oli have correctly defined the artificial as an object obtained by means of technical expedients or procedures which *imitate* or replace the appearance, the product or the natural phenomenon. Likewise, the imitation component is defined by the same authors as the capacity to get or to pursue, according to some criterion, varying degrees of similarity. The ambiguity of the question emerges, however, from the definition of the adjective 'feigned' which, according to Devoto and Oli, defines a product obtained *artificially* as *imitation*.

Undoubtedly men but also many animals, are familiar with the art of imitation and of deception. (But, by the way, who would have accepted, for instance, the expression 'feigned intelligence' rather than 'artificial intelligence'?). Anyway, the semantic weight of this feature on the concept of artificial, definitely seems to be too high.

M. Negrotti, *The Reality of the Artificial*, Studies in Applied Philosophy, Epistemology and Rational Ethics 4, DOI: 10.1007/978-3-642-29679-6_2,
© Springer-Verlag Berlin Heidelberg 2012

The *perspectiva artificialis* of Leon Battista Alberti and Piero della Francesca—but also the landscape paintings of the so-called Quadraturism school, born in the sixteenth century and enjoying success in subsequent periods (for instance, with Andrea Pozzo and his vault in Saint Ignazio in Rome), may be defined 'feigned', if you will, but only in the sense of something modeled, moulded by man as it is for the Latin origin of the verb *fingere*. However, the proper meaning of the term artificial is something that has often circulated in our culture in the most different fields, for instance in the following statement by Thomas Jefferson

> I agree with you that there is a natural aristocracy among men. The grounds of this are virtue and talents there is also an artificial aristocracy, founded on wealth and birth, without either virtue or talents; for with these it would belong to the first class. The natural aristocracy I consider as the most precious gift of nature, for the instruction, the trusts, and government of society. May we not even say, that that form of government is the best, which provides the most effectually for a pure selection of these natural *aristoi* into the offices of government? (Jefferson 1813)

To sum up, though in every naturoid there is a deceptive or 'illusory' component by definition, it is not its only component, because its main character is, rather, to reproduce something natural. While we may deceive people through various kinds of technological devices, such as conjurers and illusionists, but even designers and architects do, the deception coming from an artificial device is the most impressive since through it humans savor the power not only to dominate nature, but also to rebuild it at will. Thus, as described by Pliny the Elder, in the competition between the painters Zeus and Parrasio, the former was so skillful in drawing bunches of grapes that the birds themselves were attracted to them; the latter, in turn, drew a sheet which seemed to cover a painting so realistically that Zeus himself was deceived by it.

Likewise, as described by Nicholas Negroponte, adding realism to an artificial system may sometimes have very strong effects on man too. When in the 1970s one of the first teleconferencing systems was designed in order to make the emergency procedures of the US Government more efficient, a device was added to it, by means of which a moving plastic head indicated the person who was speaking at every moment, for instance the President. The result was that

> … video recordings generated this way gave a so realistic reproduction of the reality that an admiral told me that those talking heads gave him nightmares (Negroponte 1995)

In the above-mentioned examples and, overall, in the great intellectual achievements in painting during the Renaissance, it is very clear that the deceptive and 'illusory' component of the artificially reproduced things, i.e., 'fiction', is generated at different levels and with seemingly diversified meanings. Indeed, while in the case of painting the fiction is an intrinsic aspect of the object, in the case of the reproduction of the President's head cited by Negroponte, it is a secondary feature of the audio reproduction of his presence.

The famous desperate plea of that great sculptor who turned to his work and asked "Why do not you speak?" reinforces this point and the definition itself of naturoid we are proposing here. Actually, a naturoid, as an artificial attempt to reproduce nature "using different means", seeks similarity, and, if it succeeds in achieving likelihood, deceives precisely because it is a matter of similarity and not of identity. Nevertheless, what is important is not the deception but, indeed, the accuracy of the reproduction in the eyes of those who have to use or adopt the naturoid.

In this sense, as affirmed Prof. Willelm Kolff—one of the most important artificialists of the past century, who designed the first artificial kidney during World War II, and who then worked in the field of the artificial hearts—an artificial heart tends to 'cheat nature' because the blood it pumps arrives to the concerned organs 'as if' it had arrived from the natural heart. Nobody, anyway, would reduce such a device to a 'feigned' heart.

The fact is that the deception to which common language refers is usually associated with artificial reproductions due to some of their external or 'aesthetical' appearances, such as the aspect we can perceive in theatrical scenery, the crying of a traditional doll, the appearance of an architectural remaking by Ulisses Morato de Andrade or—but with greater caution because we are dealing with great art—the artificial perspective of a painting by Piero della Francesca.

In conclusion, the term 'artificial' always implies the work of man. His 'art' in the broadest sense and the result cannot, therefore, but show traces of its origin: not nature but technology, even here, in the broadest sense of the word.

In this frame, the term 'naturoid' includes only a part of the artificial domain, in its broad sense of something 'man-made'. The fact to be artificial, i.e., 'man-made', indeed, is only the *necessary* condition for having to deal with naturoids. In order to be true naturoid, an object has to satisfy the second condition, namely a *sufficient* one: it must be designed to reproduce an object existing in nature.

Even the definition of artificial as something which is set against the natural—as was maintained by the Hellenistic culture—is being called into question here, because it surely does not apply to the special artificialistic area of naturoids. How would it be possible for the blood pumped by an artificial heart to be used effectively by a natural organism if it came from an object set against nature?

The true opposite of the natural is the conventional artifact, i.e., the product of conventional technology which, both in terms of material and, above all, in terms of its functions, leaves nature out. Actually, an object produced by conventional technology exploits some laws of nature, is subject to natural constraints and acts on the natural world intentionally trying to change it. It is sometimes beneficial and sometimes creates problems of various kinds, but those problems are different from those caused autonomously by nature.

A naturoid, on the contrary, cannot exist without something natural that it refers to and it tries to reproduce. Said differently, a naturoid has a sort of umbilical cord which links it to nature and, at least in its first stages of development, it does not attempt to change it. In fact, at least ideally, an artificial heart would leave the organism and its functioning unchanged. But, what really happens is that the ability of the technology of naturoids to increase natural heterogeneity, triggering something unexpected, depends on its conventional technology content. Recently, it has been underlined that one of the most peculiar features of human technological activity is

> … an ability to engage in what can be called 'second-order instrumental action'—that is, making an artefact that only has utility in making, or by serving as a functional part of, another artefact (Aunger 2009)

The interesting thing is that, in making naturoids, humans are engaged in the very intriguing adventure of designing and building *second-order* devices that will be able to interact with 'machines' designed by nature and its evolution and not by themselves. Therefore, if the device will not work well, then humans cannot but change the design of the naturoid, without stating that the project of the whole 'machine' is not feasible, because it has existed and worked for a very long time in nature.

Chapter 3
Duplicating Reality

Whoever has some familiarity with an electronic copier knows exactly what is meant with the term 'copy': the reproduction of a document or an image onto another sheet of paper. The copy may be black and white or color, but, in any case, it is nothing but a photograph, at a given resolution, of the original document.

In contrast, if one does not have such a machine or a camera at his disposal, but only a sheet of paper and a pencil, then he could 'copy' the original text, summarize it or sketch the image just to render the idea. If he had neither a sheet of paper nor pencil, he could only try to memorize the relevant points of the text in question or the main features of the image.

In all three of the above-mentioned cases, the original remains what it was before it was copied, summarized, and so on. The only new reality will be the reality of the copy, our sketch or memory. Furthermore, the new reality resulting from our action will reflect the materials and procedures we have adopted. The resolution of the copier, and its ink, will modify the aspect of the original document or image in some measure, particularly if it is colored ink, while our style and our choices in summarizing a text or in sketching an image, might even distort them.

In the third case, the 'materials and the procedures' we adopt, will be consistent with our perceptions, our memorization abilities, and related biases or habits we have acquired in our experience. Nevertheless, if the original text were a mathematical or chemical formula, or an analytical report of an event—i.e., pure *information*—the accuracy of the copy, in terms of colors or drawing style, would be of little importance. As in the case of a train timetable, which could be written in very large characters on a wall or in very small ones in a brochure, the important thing is that the original information is rendered accurately and in its entirety. On the contrary, if our interest were just in the original as such, for instance in its graphic or aesthetic style, then its information content would be less important. We could be tormented by the obsession, we see in collectors, to possess the original for its unique value, and no copy would give us sufficient satisfaction.

M. Negrotti, *The Reality of the Artificial*, Studies in Applied Philosophy, Epistemology and Rational Ethics 4, DOI: 10.1007/978-3-642-29679-6_3,
© Springer-Verlag Berlin Heidelberg 2012

In a word, man can only make copies of reality—which are identical to the original, i.e., duplicates or replications—if he is dealing with informational realities, such as computer programs, but not concrete objects. Mass produced technological objects, where the prototype is reproduced by means of the same materials and procedures, constitute exceptions which imply not natural exemplars, but rather artifacts or systems intentionally and even formally designed by man himself. In the area of natural concrete objects—constituted by matter organized in a given way—replication is possible only through natural means and, of course, only where this is done by nature itself, as in the case of biological cloning controlled by DNA.

In every other area, man can only resort to the artificial reproduction. A naturoid is not, therefore, a replication of reality, but a reproduction, i.e., a production based on a natural *exemplar,* using materials and procedures different from those adopted by nature. It must be noted that the constraints due to the materials and procedures are unavoidable. Indeed, reproducing a natural object, e.g., a flower, using the same materials and procedures, would mean replicating it, and this, as we have seen, is possible only by means of a cloning process, and, this way, we would place ourselves outside the domain of naturoids. Up to now, Frankenstein belongs to the realm of fantasy and not to the realm of bioengineering.

However, this does not mean, that man is unable to modify nature acting on its own elements. As in the case of genetic manipulation, to cite just one example—but even in the more traditional procedures of crossbreeding animal or vegetable species—the recombination of fundamental structures of life is now within our capabilities.

Nevertheless, all this has nothing to do with the realm of naturoids because, in their building, man acts within nature, distributing, as it were, different playing cards, giving birth to non-biological systems not included in natural evolution.

In this sense, the understanding of the artificial content of the term naturoid, we introduce here, must be distinguished from the definition which has been widely accepted since Lucretius, according to which everything is 'artifice', because nature itself is able to make its own modifications, including those carried out by man. On the other hand, more recently, the chemist Roald Hoffmann, made a very persuasive argument that even so-called natural cotton does not differ so much from other synthetic fibers.

> A typical field of Egyptian cotton receives several treatments with insecticides, herbicides, and chemical fertilizers. The fiber is separated from the seed (ginned), carded, spun into a yarn. For modern shirting, cotton is also treated in a variety of chemical baths, bleached, dyed. It may be 'mercerized', strengthened by treatment with lye (sodium hydroxide). Optical brighteners or flame retardants might be added. Eventually the cotton is woven into cloth, cut, and sewn into a garment. It may be blended with another fiber for strength, comfort, or some other desirable property. That's an awful lot of manipulation by human beings and their tools, and to sharpen the point, manipulation by *chemicals*, synthetic and natural, going into your *natural* cotton shirt! (Hoffmann 1997)

These intellectual positions are quite convincing, but they are also largely reversible since, based on the same line of reductionist reasoning, one could maintain that everything is natural because everything is made of atoms and molecules. Such premises are of little use if one is aiming to gain an understanding of the possibilities, limits and consequences of human attempts to reproduce what humans observe in the natural world; i.e., the results of variably complex combinations of atoms and molecules already generated, in given ways, by nature over millions of years.

A naturoid, according to our definition, which is consistent with a part of the historical use of the term 'artificial', consists in the result of human efforts to achieve the same results as nature, using strategies which are different from those employed by nature, and, therefore, *lato sensu* technological. All this has nothing to do with, nor does it contradict, the thesis which maintains, *at a deeper level of analysis*, that all that happens in the universe is, by definition, internal and, therefore, natural (artificial included). Rather, even from this standpoint, it is possible to deduce that the realm of naturoids, though it is related to nature and without referring to nature it would make no sense, has the intrinsic outcome, although unintentional, to set up a new reality, a third technological reality inspired by nature.

In conclusion, it is reasonable to maintain that a naturoid is always related to something which it is not, and from which it draws its *raison d'être*. While nature is what it is, as it were, in absolute terms and conventional technology creates artifacts or processes compatible to a greater or lesser extent with nature but, in any case, not present in it, the technology of naturoids generates objects, processes or machines surely technological, but 'suggested' by nature. Therefore, at least ideally, a naturoid should exhibit features which are not only compatible with nature, but which would pretend to overlap nature.

On the other hand, a true general technology of naturoids, as an autonomous body of knowledge, techniques and materials, clearly does not exist, nor is it sure it could exist. Actually, intrinsically artificial materials or techniques, able to be applied to any naturoid project, do not exist: rather, there are and there will be natural materials recombined by man using techniques derived from conventional technology and, let us say, forced to reproduce some natural object or process. In other words, the designer of naturoids can only exploit materials and procedures made available to him by conventional technology. This is true for Icarus and the wax he used to fix his wings to his body and for the alimentary canal of the famous duck made by the eighteenth century mechanical engineer Jacques Vaucanson. Vaucanson was immediately interested when he heard about a gum from the Indies, since it sounded to him like an ideal material to reproduce the internal tissues of his artificial animals (Fryer and Marshall 1979; Bedini 1964).

The same holds true for today's bioengineers who pay a great deal of attention to the discoveries made in the field of materials technology and in every other conventional technology, in order to find the most suitable component for their projects involving artificial organic structures or processes. The designers of naturoids have always adopted materials and techniques developed by conventional technology, even making some specific requests to technologists, in order to reproduce natural objects or processes. The world of design resorts to a huge exchange of materials and components that flow from one field to another according to the needs of projects. In this area, the technology of naturoids acts as a filter because its needs are not only due to the design preferences of the man because, first of all, they are already established by a very special customer, i.e., the nature. Anyway, just as technologists of naturoids always resort to conventional technology in order to realize their projects, conventional technologists resort to the artificial in order to develop or improve objects, processes or machines that, as such, do not have reproduction aims. One of the most relevant example in this direction could be that of cybernetics, whose basic models, which govern the self-regulation of biological systems, have been widely applied to conventional technology devices, like cars and aircrafts, computers and military weapons and several other technological systems.

Thus, as far as the technology of naturoids is concerned, we are facing a sort of paradox. On the one hand, a naturoid is intentionally related to nature since it aims to reproduce it. On the other hand, it inevitably depends on conventional technology, i.e., on a technology that, as we have seen, does not establish any reproduction target but, rather, builds objects, processes and machines which are heterogeneous compared to nature. As a consequence, a naturoid swings between nature and conventional technology completely overlapping neither the former nor the latter. In fact, if it would overlap nature, it would coincide with natural replication and, if it would overlap conventional technology, it would distance itself inexorably from nature.

The destiny of naturoids, in whatever field they appear, cannot but bear the marks of this paradoxical ambiguity which, as we shall see, is the primary logical cause of its tendency to establish itself as a *sui generis* reality.

Part II

Chapter 4
The First Step: Observation

Whoever wants to design a naturoid, i.e., a man—engineer, artist or other who is attracted by the idea of reproducing something natural—is strongly characterized by a special way of viewing the world.

First of all, he must have a keen interest in observing nature, since it is from nature itself that he gets his ideas. Those who are dedicated to the design of artificial things should be able to grasp those aspects of reality which have a greater likelihood of being reproduced, just as scientists are sensitive to the perplexing and as yet unexplained aspects of what they observe, and artists, in turn, concentrate on other aspects which allow them to interpret those aspects meaningfully, that is to say through their poetics.

Conventional technologists are also closely related to nature and, therefore, dedicated to its observation, but their main aim is to design objects, processes or machines which are able to control or modify natural events and not reproduce them. Conventional technologists often see nature as a problem or an adversary, while the designers of naturoids look at nature as a source of project to realize.

The four figures we have sketched—namely scientist, artist, conventional technology engineer and designer of naturoids—resort to four types or 'styles' of observation which are found, in varying degrees, in every human being, but are extremely relevant, above all, in the above-mentioned professions.

These types of observation are often interlaced—as in the case of Leonardo da Vinci—but they possess their own qualities and features. For example, a major advance in the development of the microscope was made thanks to a casual observation made in the seventeenth century, when, in order to enhance its magnifying power, the spherical convex shape of a drop of water was used as a model. This discovery had already been made, without any subsequent development, in ancient times when it seems that spherical bottles full of water were used as magnifying devices. The observation of a physicist or chemist would be concentrated, of course, on the way a drop of water forms, stabilizes morphologically, or on its internal and external dynamics, while for those who were interested in

M. Negrotti, *The Reality of the Artificial*, Studies in Applied Philosophy, Epistemology and Rational Ethics 4, DOI: 10.1007/978-3-642-29679-6_4,
© Springer-Verlag Berlin Heidelberg 2012

improving the microscope, the drop was relevant as something natural to be reproduced. Thus, innovators of the microscope are true designers of naturoids, as was Watson-Watt, the inventor of radar, who, is said to have gotten the idea for his invention observing the way bats detect obstacles during their flight.

In any case, it is clear that, to reproduce something, first of all one must observe that thing and, as a consequence, observation is the first, unavoidable step every naturoids maker must take.

But what exactly does observation mean? Discussing the observation process would lead us astray down the often difficult paths of philosophy, but the problem in itself is not difficult to understand, and the solution we propose is simple, at least within this work.

As far as the observation process is concerned, the delicate point lies in its unavoidably selective status. For instance, everyone can observe the moon, but no one can observe it in its entirety; or to cite a more relevant example of the relative nature of the observation process, consider the observation of a landscape. Clearly a geologist will observe a landscape in a different way than an agronomist, a botanist or a painter.

In this work, we shall call the profile we observe reality from *observation level*.

This concept bears some similarity to that of *levels of analysis* or to that of *description levels* adopted in some theories of complexity and to the *reality levels* outlined by Oppenheim and Putnam (1958). It also recalls, to some extent, the *observation perspectives* of C. Morris's 'objective relativism' when he maintains, although in a metaphysical scheme, that something exists only

> … within a perspective, and there is no all-embracing absolute perspective within which everything exists. (Morris 1948)

An analogous position, on a scientific epistemology plane, is implicit in the thesis put forward by Kuhn regarding the ability of scientific paradigms, during the phases of 'normality', to direct the observer mainly towards that which is consistent with the dominant paradigm.

The preference for the term 'observation level' arises from two factors. First of all, observation levels cannot be enumerated and placed in any arbitrary taxonomy. They have to be considered as infinite in number as the possible positions we may place ourselves in the space or within whatever descriptive dimension. Second, the concept of observation level, underlying the selective, and sometimes constructive, character of human observation, characterizes not only scientific exploration, but also any sensible interaction we may have with the empirical world, and, subsequently, any description of an observed object. We may describe an object at more than one observation level, but we can adopt only one such level at any given moment. The sciences themselves constitute, for all practical purposes, the institutionalization of the observation levels thus far discovered or constructed by scientific reasoning. In fact, there is a closed analogy between the constraints regarding the combination of observation levels in the field of naturoids—that is to say when designers want to reproduce a natural object selecting two or more observation levels—and the cooperation among different scientific disciplines.

Indeed, what we call 'interdisciplinary', when it succeeds, consists of the setting up of a new observation level rather than a 'sum' or a generic 'synthesis' of two or more levels.

Biophysics or biochemistry are good examples of this kind. Combining knowledge, lexicons and techniques of the disciplines involved, they both give rise to largely autonomous new sciences, no longer easily comparable in terms of knowledge, lexicons and techniques with the scientific fields they come from. Normally, from biophysics research we cannot expect new knowledge of basic physics nor of a general biological type, but, rather, findings concerning those aspects of biology which are of an exquisitely physical relevance, such as the effects of radiation on living systems, the transmission of nervous pulses, and muscle contraction.

When an interdisciplinary project does not succeed, one of the observation levels involved dominates the others. This may occur for several possible reasons, including the greater development of one of the disciplines compared to the others. It is a matter of a destiny which also seems to be true for some disciplines in the family of bionics, dominated, as we have said, by the temptation to reduce the most heterogeneous phenomena to a pure informational level.

It is clear that the problem of relativity of our observation ability is of particular significance not only for philosophers, but also for scientists, and, in a very particular way, for designers of naturoids. When, for example, we design a human skeleton as an aid for teaching anatomy, which observation level should we adopt? Usually, a plastic skeleton will not encompass the molecular level nor the atomic level but will orient itself at the typical level of macroscopic anatomy and physiology. In other words, if we dissected the bones of an artificial skeleton, we would not be able to observe the biological structures that constitute bones in nature. On the other hand, even if an anatomy institute were very demanding and required that even the biological structures of bones be reproduced, using, however, materials which are different from actual bones, a threshold would soon be reached, beyond which one could not proceed, because of both our lack of knowledge and the difficulties involved in realistically rebuilding the connections between the various levels. This explains why any artificial representation of the human body, be it a two-dimensional one, like a drawing, or a three-dimensional device, like a skeleton, always comes from a selection of a profile, or an observation level, of the whole organism whose 'synthesis' cannot be represented at all. A simple look at Leonardo's representations of the human body anatomy, drawn between 1478 and 1518, will well clarify this point. This is true for all human concrete representations of things, of course, and this fact clearly illustrates what we mean by concepts such as 'analytical models' of something, as compared with the wholeness of that thing, whose synthesis cannot be but a concept and not a workable representation.

Also the painting by Jan Brueghel the Elder, *The Entry of the Animals into Noah's Ark*, 1613, explains what an observation level is better than a long discussion. The traditional representation of Noah's ark may help us to understand how man is naturally oriented to perceiving reality in anthropomorphic terms: all the animals which were selected to survive have physical dimensions which are compatible with our senses, while microbiological living systems are damned to extinction.

In the field of artificial intelligence, no one would expect a computer program able to reproduce the logical intelligence of a human being (e.g., the ability to perform correct deductions) to suddenly exhibit the capacity to analyze the meter of a piece of poetry. The same is true for artificial intelligence programs able to recognize geometrical forms or sound samples: the computer can easily shift from one target to the other, but if it has to recognize a musical instrument taking simultaneously into account both its form and its sound, then it would need to work at a higher and new observation level.

In short, even intelligence can be observed, and therefore, described and modeled at different analytical levels, and in order to take more than one into account simultaneously, we need to enter a new observation level.

For the time being, we shall limit ourselves to asserting that human beings are forced, by their own nature, to consider only one observation level at a time. We can rapidly shift from one level to another, but each time we make a shift, we change radically the content of our observations, descriptions and judgments regarding what we observe. As we all know, a clear day can be cause for happiness or, in contrast, a real problem depending on the activity we had planned for that day (an excursion in the mountains or some experiment with a new kind of rain gauge), though it is, in itself, the same climatic circumstance.

Chapter 5
Observation and Representations

The limitation of our observation capacities to only one level at a time means that the naturoid which is made at the end of the design and building process in no way will be the reproduction of the exemplar in its entirety, that is to say an object that could be identified with the natural instance at *any* observation level. This would be true even if we could use the same materials and procedures used by nature. Though this case lies outside the field of the artificial properly defined, even in the above circumstance we would be forced to use only those natural materials we have observed and not, of course, those remaining hidden and detectable only at other levels. This is a well-known fact for all those who have tried to reproduce some kind of fruit, flower or even some of their alimentary derivatives, using the same seeds and the same procedures nature does, but neglecting, or completely ignoring at all, other components—such as the composition of the air or the climatic dynamics—which make the development or the production of the exemplars possible.

As we have already noted, the need to use materials and procedures that are different from those used by natural exemplars, introduces a decisive constraint which prevents the naturoid from approaching the exemplars beyond a certain threshold.

Nevertheless, we must again clarify another point concerning the observation process. Very often, we argue that, in the end, we see what we 'want to see', rather than some objective reality. This is an issue still well known to modern scientists—though in naïve terms, but relevant for the history of science—in their anxious, and often frustrating, search for instruments and procedures for surveying reality 'as it is'. Just consider the following quotation of the founder of histology, the French scientist François Bichat, who, concerning the observation made possible by the microscope, said:

> …it seems to me that this instrument is not of great use to us, because when men look in the dark, everybody sees in his own way … (Galloni 1993).

M. Negrotti, *The Reality of the Artificial*, Studies in Applied Philosophy, Epistemology and Rational Ethics 4, DOI: 10.1007/978-3-642-29679-6_5,
© Springer-Verlag Berlin Heidelberg 2012

Today, many philosophical doctrines focus on the process of observation in science, pushing convictions that have regularly appeared in the history of thought for the past two thousands year to the extreme. In these doctrines reality does not exist at all, or, better yet, its existence has no objective relevance at the moment we observe it. What is truly important, according to these doctrines, is the act of 'constructing' the world, independently of its objective reality. This is the position, for instance, of Bateson when he maintains that the importance of the reality

> must be denied not only to the sound of the tree which falls unheard in the forest but also to the chair that I can see and on which I may sit. (Bateson 1972)

If we completely accept such a position, then we should deny the objective validity not only of our daily observations, but also of those made by scientists: everything would be uncertain, subjective and incomprehensible. On the contrary, it is clear that, at least in the field of natural phenomena, i.e., those phenomena that can be precisely, though conventionally, measured using instruments, the world has its own objective capacity to act on man, starting with its actions on our senses—just consider an earthquake. In this case, the epistemological position of the observer in no longer relevant.

It is clear that the designers of naturoids do not accept the constructivist thesis and strongly believe in the existence of natural objects, because, without this belief, they would not be able to observe, and then reproduce them. In other words, the science and technology of naturoids share a community-based agreement on what reality is, at least in pragmatic terms. As the history of science clearly shows, scientists, like other people, often are wrong in attributing to natural objects some kinds of features or functions because of the limitations of some observation level, and as a consequence, of any representation or model. Nevertheless, the benchmark of experimental work allows them to permanently correct their empirically based views.

If we referred to the constructivist doctrines, it is only because they, at least, remind us of the fact that nature and its events, as we have already underlined, are not at our disposal in an 'immediate' way. Our knowledge of the world is 'mediated' by our mind, the place where we form *representations* of reality. A representation is the mental reproduction of what we observe through our senses or something we generate autonomously, for instance the image of a face or lake, but also our subjective description of the atom or of a continent, and even our 'vision' of the universe or God. Therefore, a representation is something 'meta-artificial', a sort of prelude, as it were, to the design of a naturoid: it is, in fact, a non-material construct which reproduces the world we observe and is definitely useful for surviving in it. Without representations we would not be able to evaluate situations or associate memories and observations: we could only rely, each time, on immediate reactions to reality as many animals surely do. Fire would burn us every time, because, within our mind, we would have no recorded notion of its features and effects. In practice, when we encountered fire, we would not recognize it, and, likewise, its symbol drawn on a wall—as a result of a collective representation—would have no meaning for us and we would not be kept on alert.

In short, the representational system adopted by the subject gives meaning to his action and this meaning may evolve during the problem solving process (Leiser et al. 1976).

Forming representations is closely related to our perception and observation, and different classes of representations correspond to each level, allowing us to adapt our behavior to different circumstances. The same is valid in the field of naturoids, where, according to different objectives, a human heart can be represented, modeled, and then reproduced, in a very wide range of versions, namely aesthetic, instructional, or as a project of bioengineering.

The mind does not play a passive role in forming representations because, as in the choice of an observation level, the whole system of our experiences and our preferences, interests and fears, acts on it. Culture itself is a powerful source of 'directions' we should follow, or refuse to follow, in observing reality and, therefore, in forming representations. For instance, the sociological and the psychological dimensions of man are today widely accepted as a reality, but it is a matter of a relatively recent fact. Of course, man has always had what today we call a sociological or psychological dimensions of life, but the related representations, models and theories, specifically dedicated to these phenomena, did not appear until the nineteenth century. This scientific void prevented these dimensions from becoming observation levels, together with daily, political, or spiritual observation levels.

The same holds true for other observation levels: the subatomic or the ecological, the economic, the magnetic, the micro-biological or the chemical, and so on. Each of these levels is, simultaneously, the cause and effect of representations which, almost referring to the same natural object or process, privilege one profile only, only one way of being and presenting themselves to the observer.

Drawing further upon Heisenberg, we can speak of a sort of generalized indetermination principle: the choice or the construction of an observation level allows us to capture a true reality, but only the one which is compatible with that level. In more general terms, one could say that in selecting an observation level,

> … we force the matter to choose a configuration from among those that are available … (Regge 1994)

The constructive bias of our mind, rather, is an important issue in the field of the art, where the artist's 'views' of the world have no empirical benchmark to comply with and, therefore, the value of his 'interpretation' of a natural object or event does not depend on the public assessment of its objectivity.

Perhaps it is why Oscar Wilde maintained that art is nothing but a form of emphasis, exaggeration of the reality we perceive. Though, since Aristotle, art has been assigned the role of imitating nature, *de facto* no artist can generate nor usually does he wish to generate copies of what he observes, but, rather, an interpretation of it according to his own poetics, i.e., his own representational modalities, or those of the school he belongs to.

The history of painting, for instance, exhibits a very wide range of pictorial observation levels of representations that are very different from each other, even when they deal with the same 'subject': from the spirituality of the Middle Ages—where what the properties of the sacred images should be were imposed by the official doctrine of the Church—to the concreteness of the Renaissance, from the vagueness of Impressionism to the complexity, often solipsistic, of the vanguards (Bertasio 1996). Indeed, it is through art that we can understand to what extent mind and culture may sometimes be active protagonists in forming or in confirming many orders of representations. Just consider painting in the Middle Ages, the pictorial reproductions of God or the devil—for instance, the chilling reproduction of the devil by Coppo di Marcovaldo in Florence. This demonstrates that, at one extreme, man is able to generate artificial objects based on exemplars which are quite inexistent in nature, but, nonetheless, accurately represented in cultural traditions. As we will see in Appendix A, artists build up representations at their own peculiar observation levels which do not come from some conventional agreement within a community, as they do in the case of scientists, technologists and ordinary people.

Likewise, the current representation of the atom, introduced in physics by Niels Bohr in 1915, with its analogy with the solar system, has become a true graphic symbol widely shared by scientists and common people, though it is very far from being verified in its structure, dimensions and dynamics. However, it is extremely useful and necessary and makes the research in the field of physics possible. In general terms, observation levels depend on personal or cultural premises that orient our way of looking at the world. Thus, the overall feasibility of naturoids will be evaluated based on the plausibility of our description of the object or the process we want to reproduce. For instance, in the field of nanotechnology, if we look at natural things and processes—life included—as machines made of molecules, then, as Bensaude-Vincent has rather ironically remarked,

> through a continuous process of mutual transfer of concepts and images

between biology and materials science, surely we can arrive at

> … a common paradigm based on an artificialist view of nature. Nature is populated with nanomachines that human technology should be able to mimic or even to surpass (Bensaude-Vincent 2004)

The fact is that if a design fails in its goal it will be difficult to establish if the failure is merely due to bad design work or to a fallacy of the observational premises it comes from.

Chapter 6
The Exemplar and Its Definition

As we said, the exemplar is the natural object or process which is chosen as the target of the reproduction. More precisely, we should say that a naturoid is the reproduction of the representation of the exemplar, which the artificialist generated in his own mind. Models, even purely mental ones—in the technological design as well as in art—are examples of formalized representations which function as pilot maps or schemes of the natural object or process which we wish to reproduce.

The choice of the exemplar, however, is not a simple task, devoid of ambiguities as it might first appear. We all know that, in considering an artificial heart, we are referring to a well-known and recognizable exemplar, that would appear to be easily distinguishable from all that is not a heart. Obviously, for an engineer the question is much more complex: which organic parts, vessels, muscles, sub-systems, define the heart? In other words, what are the 'boundaries' of a heart?

In addition to realizing heart valves, today there are also devices which aim to re-create the left ventricle (the so-called *left ventricular assist systems*) and are designed to work together with the natural heart of the patient. Other devices reproduce both ventricles. The total artificial heart, able to completely replace the natural heart, has only recently become an achievable aim, but many problems remain. Often these problems are related to the dedicated sub-system—which, therefore, must be considered as a part of the exemplar—generating electrical power for the control of the electronic circuits and moving parts.

In another field, suppose we wished to reproduce a natural pond, seen during travel in the UK, in a meadow surrounding a house under construction in Tuscany. How should we establish its boundaries? On a topological level, should we even include the geological structure the pond's bottom and sides or not? As far as the flora and fauna are concerned, how close should we come to the natural pond? Should we include every aspect of the pond's ecosystem, which ranges from ducks and fish to microbiological creatures? It is quite clear that, different answers to such questions—not to mention the climate context requirements—will give rise to

M. Negrotti, *The Reality of the Artificial*, Studies in Applied Philosophy, Epistemology and Rational Ethics 4, DOI: 10.1007/978-3-642-29679-6_6,

© Springer-Verlag Berlin Heidelberg 2012

different models and concrete reproductions of the pond, and these differences may be critical, if related to our aims.

The same is valid for artificial lakes, because they

> … are typically much shallower than natural lakes—explained Charles Goldman, professor of limnology at University of California, Davis in the department of environmental science and policy, and director of the Lake Tahoe Research Group … They're often so shallow that they do not stratify, or circulate from top to bottom … Which means they're excessively productive—green, essentially. Artificial lakes are particularly hard to manage … like an algal generator. This lack of stratification, along with the fact that typically, such lakes are surrounded by fertilized lawns or function as drains for storm waters—causing excess nutrients to flow into them—can result in eutrophication (Goldman 1999)

In the field of artificial intelligence this is a well-known much debated problem: how can we define—a verb which, by the way, derives from the Latin *definire*, which means 'to fix the boundaries'—human intelligence in comparison with other functions of the mind like memory or intuition, fantasy or curiosity?

At one extreme, we could consider exemplars from the animal field, such as the *holothuria* (it is also known as the sea cucumber) which lives symbiotically with the little fish *Fierasfer acus*. To what extent is it correct to separate these two *symbiotic* entities, first of all in representational or modelling terms, and then in terms of design and reproduction? This fact is rather widespread in nature and includes several insects who survive through symbiosis with particular trees, like hornets with fig trees (Dawkins 1996).

It is quite clear that the task of outlining an exemplar consists in an operation which is always arbitrary to some extent: a true isolation of an object or process from a wider context, which includes it or from an environment which hosts it. Moreover, if the exemplar is a biological system—whose linkages with the environment are strategic not only for its survival, but also for its regular behavior—then the arbitrariness of fixing the boundaries is very crucial. A mistake in this direction could result in the failure of the project or, alternately, lead to something quite unexpected as many times happens in the field of naturoids.

Western civilization, because of its philosophical and scientific tradition, has demonstrated an ability to carry out 'analyses' of the natural world, greatly benefiting from such analyses. But analysis—a term which, interestingly, derives from the ancient Greek 'to break down'—has always some costs in terms of the effectiveness of the scientific explanation and prediction, and, more, the design of a naturoid. In fact, an incorrect scientific hypothesis will simply bear to some weakness or failure of an experimental prediction. For instance, in the nineteenth century, the results of experimental thyroidectomies to elucidate the exact functions of the thyroid gland were misinterpreted because investigators also inadvertently removed the parathyroid glands (Hamdy 2002).

For its part, an incorrect reproduction of an exemplar, due to either an incorrect knowledge of it or a bad fixing of the boundaries, will lead not only to the failure of the reproduction, but very often to unexpected behavior of the device.

To address this will require, in turn, more than one observation level, and the analysis, with its usual isolation strategies, might not be able to make all the required levels observable. This deficiency, in turn, will propagate to the fixing of boundaries of the exemplar.

In other words, the choice of an exemplar is a sort of literal 'eradication' of some part of nature and this can take place, as we saw, both in terms of a true isolation in space, and in structural terms. On the other hand, man seems to have no alternatives: as we see in the simplest daily observations, or even in observations aimed at defining an exemplar to reproduce, man cannot but proceed by putting one thing at a time in the foreground and relegating to the residual things to the background. The process of reducing the complexity of the exemplar involves also the forming of representations and the arrangement of a model, of course.

All this happens as a function of the choices imposed by observation levels. Hence, the functioning of a naturoid will be similar to that of the exemplar only to the extent that the observer, or whoever has to use it, evaluates it by placing himself at a level which is as close as possible to the level of the designer.

Of course, if the observer and the user have to deal with an artificial device conceived and built based on an observation level that has never been experienced or that is extremely subjective, then they will have to face many additional problems in evaluating the quality of the reproduction. This is true not only for artistic reproduction—which is innovative by definition—but also for scientific innovation process. Actually, the reproduction of the solar system according to a Copernican representation, by means of a mechanical device, would not have been easily understood and appreciated at a time when the Aristotelian representation of the universe was the commonly shared model. On the contrary, under the influence of the mechanistic culture of the sixteenth and seventeenth centuries, Kepler was able to develop a three-dimensional reproduction of the universe, the famous *Machina mundi artificialis*, which was subsequently given to Prince Friedrich von Württemberg.

Thus, defining the exemplar is a task which, though it appears to be quite obvious in daily life when we indicate the objects or the events that surround us—or happen within us—proves to be quite complex as soon as we try to begin a project or design intended to reproduce it.

One of the lessons to be learned from the study of naturoids is our own limitations, first of all, in knowing nature, which is a phase that precedes—or should precede—the phase of reproduction.

Analysis often allows us to control reality, such as when, for instance, we succeed in describing with some accuracy the anatomy, physiology, and perhaps even the psychology of a given animal, e.g., a pigeon. Thus, we can explain many of its behaviors, pathologies or abilities. This knowledge, however, is not sufficient for designing an artificial pigeon, not only because it is only a fraction of what we should know about a pigeon, but also because it has been acquired, unavoidably, through analytical strategies. These strategies, break reality into areas which are often as heterogeneous as the observation levels they derive from, and their synthesis would require knowledge that could only be acquired through as yet unknown strategies.

Chapter 7
Essentiality of Things

The choice of an observation level and an exemplar are the first two steps in the process which ultimately leads to the design of an artificial object.

Nevertheless, the choice of an exemplar is not the final conclusive defining moment of what will be done. As soon as the problem of the delimitation of the exemplar—which we discussed in the previous section—is solved, we are faced with a new and decisive step. This moment could be defined as the choice, or the attribution, of an *essential performance* for the exemplar.

Essential performance is the quality, function or behavior of an exemplar or even simply the aspect of the exemplar that the designer believes to be particular or typical, and which cannot be omitted.

In the previously mentioned case of the heart, there is a general consensus on what its essential performance is, i.e., the pumping of blood. This performance of the natural heart is fundamental: a design of an artificial heart cannot neglect such a feature. Indeed, an artificial heart that, at the observation level we assumed, namely the physiological one, was not able to pump blood would be quite unrecognizable as a heart and, more importantly, would be useless.

From a bioengineering standpoint, there is general agreement on what must be considered as essential in the case of the heart. There is also a rather general consensus regarding kidneys, lungs, pancreas, liver, skin, certain tissues, and so on. However, this kind of accord is not found in all areas of naturoids.

In the field of artificial intelligence, for instance, since researchers have to deal with a rather unknown exemplar—the human mind—there are a number of different opinions regarding its essential performances, and these viewpoints are often incompatible with one another. Furthermore, while intelligence could be interpreted as the essential performance of the mind, the mind itself could be interpreted as the essential performance of the brain, which is the sole empirically detectable object within this frame. In fact, two main schools of thought face this problem. On the one side, monists think that only the brain exists and that the mind, from a scientific viewpoint, is a useless concept; on the other side, there are

M. Negrotti, *The Reality of the Artificial*, Studies in Applied Philosophy, Epistemology and Rational Ethics 4, DOI: 10.1007/978-3-642-29679-6_7,
© Springer-Verlag Berlin Heidelberg 2012

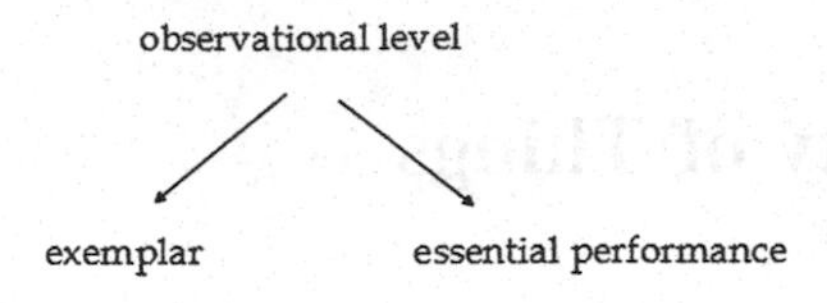

Fig. 7.1 The multiple selection model involved in the design of a naturoid

dualists who, starting from Descartes, think that the mind is a separate thing as compared to the brain, or, in a more updated view, 'emerges' from the brain, but derives its own reality through experience.

Intelligence itself, which, at first glance, might seem the essential performance of our mind, is far from being accepted by all researchers as the main mental characteristic, at least if, by intelligence, we mean the ability to solve problems. Besides other types of non-formal intelligence—such as concrete or motor intelligence—we should take into account consciousness, creativity, the ability to recognize objects or situations, memory, attention and many other functions. On the other hand, it has been widely supported that one of the most peculiar abilities of the human mind is that of finding problems; a capacity that has been highlighted since Socrates and is today conceived as very closely linked to creativity (Sternberg and Lubart 1995).

Furthermore, this issue involves the process of abduction, which is

> … a distinct form of reasoning, is the process of *inferring* certain facts and/or laws and hypotheses that render some sentences plausible and that *explain* or *discover* some (eventually new) phenomenon or observation … Therefore, creative abduction deals with the whole field of the growth of scientific knowledge. (Magnani 2007)

This means that, starting from facts, the scientist has to face a number of open branches not one of which has been already explored and which are, therefore, true sources of possible new original problems. On their part, the designers of naturoids also need creativity in their work, of course, but not in order to describe the exemplar—which is taken 'as known' enough in its essential performance. Rather, they face the problem of inventing the best technological ways to enable a device to emulate the performances of a given, natural exemplar. It is not possible to exclude that from a naturoid we could learn something more about a natural exemplar, such as human mind, but this is only a possibility. The main goal of designers is usually the reproduction of intelligence in its known—or believed—features and not finding new scientific problems, provided they don't misunderstand the functioning problems of the software or hardware device as problems of the human mind.

Problem finding, however, is a very difficult performance to be reproduced in any machine (Fig. 7.1).

Thus, it can be asserted that, in the field of artificial intelligence, there are as many schools of research or design as there are possible viewpoints concerning the essential performance of the mind, though all deal with the same exemplar.

The main schools, which have difficulty finding common ground, are the so-called schools of symbolic artificial intelligence and neural networks. The former insists on the design of computer programs that try to reproduce intelligent behavior on the basis of algorithms. These algorithms simulate mental representations and knowledge through quality and numerical symbols and through logical evaluations of these symbols (deduction, comparison, association, calculus, etc.). The latter, in contrast, draws the long-neglected cybernetic works of the 1950s, and aims to design devices whose intelligence consists in the automatic recognition—neither logical nor symbolic—on the basis of suitable 'training phases' of a network and series of diverse data: meteorological situations, geometrical shapes, several kinds of objects to be detected in ambiguous contexts.

It should be remarked that all artificial intelligence programs, if well designed, succeed in reproducing some aspects of human intelligence, but only in a stand-alone way, that is to say according to separated observation levels and related selections of the essential performances. The problem of making a machine able to shift from one performance to another, or, better, to combine or synthesize the available performances, remains beyond the current state of the art, and, as we will see, highlights a general difficulty in any field of naturoids.

The attribution of an essential performance is always a process in which empirical reality and the autonomy of the mind overlap. Attribution of a certain performance to a ductless gland may be rooted in a way of perceiving life proposed by some established theory, the premises of a religion, or other subjective preferences of the researcher. The dependence of the attribution of an essential performance on personal or cultural factors does not imply, as such, a failure in the design process, of course.

Actually, there is a strong analogy between an essential performance and the hypotheses adopted in the scientific research whose intellectual and motivational sources can frequently also be found in extra-scientific areas. Nevertheless, in the scientific domain, a hypothesis can be experimentally verified or falsified whatever its motivational source. In contrast, in the field of naturoids, the assessment of a machine will mainly depend on its ability to reproduce the essential performance attributed to the exemplar by the designer and, possibly, by the community he belongs to. The fact that the attribution may be wrong—that is to say, not objectively established by scientific research—does not imply that the naturoid will not work. If one believes that a gland has a physiological function F, and he is able to reproduce that F in an artificial device, then the enterprise could be said to be successful, even if the attribution to the gland of the function F is false. If a sort of Frankenstein wanted to reproduce a tree ignoring the light-capturing role of the leaves, attributing to them the only role of harvesting the water, then he would have built a device quite well able to collect water, but the tree would not have survived.

Really this represents a permanent potential pitfall of naturoids, especially when we have to deal with objects or systems that are largely unknown.

In every culture, the conception of man himself, is of course also based upon an inclusion of multiple essential performances. Philosophical research, regardless of the particular school of thought, is proof of the enduring effort to discover the essence of man. Especially evident is the inclusion of selections, or attributions, in the following Hindu axiomatics contained in the Chandogya Upanisad, in which one can find the

> … various steps that mark the subsequent materialization of the world: the saman is the essence of the poetical meter, the meter is the essence of the language, the language is the essence of the man, the man is the essence of the trees, the trees are the essence of the water and the water is the essence of the earth. (Schneider 1960)

Once again, in the field of biology, botany has shown that, throughout history, a strong inclination to generate attributions of changing essential performances is often dominated by extra-rational visions. We can see an example of this phenomenon in *Malva silvestris*, which, according to its classification by Linnaeus, has been used since antiquity for its medical properties, and which, according to the Pythagoreans, also boasts the capacity to save human beings from being slaves to their passions, while later, in the eighteenth century, it was appreciated for its alimentary virtues; another example is *Cheiranthus cheiri*, which, again according to its classification by Linnaeus, was used by the Greeks and the Arabs as a cleaning substance and its cardiotonic properties were only later discovered in the twentieth century.

The choice of an essential performance is a moment in which the designer of a naturoid may be as aware as he ever will be of the fact that, by privileging one performance over others—which are in principle unlimited in number—he always acts arbitrarily, and is strongly influenced, as we have already illustrated, by the observation level he assumed at the beginning of his work.

Even the above-mentioned Vaucanson was aware of this when, speaking of the digestion of his artificial duck, he defined the essential performance he wanted to privilege as follows:

> I do not claim that this should be perfect digestion, able to generate bloody and nutritional particles in order to allow the survival of the animal. I only claim to imitate the mechanics of this action in three points: in the swallowing of the wheat; in soaking, cooking or dissolving it; in allowing its going out, forcing it to visibly change its stuff (Vaucanson 1738, in Losano 1990, p. 91)

Two American researchers, M.A. Mahowald and C. Mead, among the many who are engaged in the design of an artificial retina of the human eye, made similar observations:

> In building a silicon retina, our purpose was not to reproduce the human retina to the last detail, but to get a simplified version of it which contains the minimum necessary structure required to accomplish the biological function (Mahowald and Mead 1991)

Nevertheless,

> The real vision … will probably require that artificial retinas contain 100 times the number of pixels and auxiliary circuits, to imitate the functions of perception of the movement, and to intensify the contours performed by the amacrine cells and by the ganglion cells.

> Finally, these systems will also include additional electronic circuits for recognizing configurations generated by the retina (ibidem)

Even in a therapeutic approach, in the field of optometry, there is a great need for artificial devices. Just consider treatments for dry eye syndrome, and

> … there are formulations called artificial tears that are available in the market today. They are only partly satisfactory … we need to dig in and ask what the various components of the tear fluid are, and which of the components are indispensable and we cannot do without, which are those that we can do without, but it would be nice if they were there but there is no harm if they are not there in great amounts, and whether there are components in the tears that we can do away with. The reason this question is hard is because it has been estimated that there may be a hundred different components in the human tear fluid. When we want to formulate an artificial tear fluid, we would want to do it with a number far smaller, and yet the fluid that we create should be able to satisfactorily do the job. (Balasubramanian 1998)

The same holds for the cornea and vitreous, and, unfortunately, also the Intraocular Optical Lenses exhibit their own problems, since

> A variety of optical deficits can arise following the implantation of intraocular lenses (IOLs). While artificial lenses replace cataractous lenses that have degraded markedly in opitcal quality, these IOLs can also introduce some visual side effects. These side effects include glare, halos, streaks, starbursts … (Bass et al. 2010)

Another typical case, which is analogous to the previous one, is the attempt to reproduce the propulsion used by fish with a robotic model. Until now,

> It is almost impossible to reproduce the performances of fish simply by imitating their form and function, because a vehicle able to set up uniform and continuous flexes, having a body similar to that of a fish, is quite beyond the state of art of robotics (Triantafyllou and Triantafyllou 1995)

Nevertheless, it is possible to imagine that

> In the future, such creations which are inspired by nature, will perhaps improve their biological models for some specific task, like, for instance, the exploration of the sea-bottom. (ibidem)

In its general terms, the question posed by the selection of an essential performance can be outlined in the following way, drawing from the possible design of an artificial rose. In this flower, as in every other biological system—neglecting the problems of its delimitation in the space and its structure, which we shall consider as decided—what is essential?

It is quite clear that our answer will strongly depend on the observation level we assume. The physical observation level that we have established (chosen or constructed) in order to indicate the flower as an exemplar (e.g., micro, macro or some intermediate position) will lead us to attribute some essential performance to the flower (e.g., shape, color, fragrance, some kind of behavior) and to neglect other possibilities (e.g., the consistency of the tissues, the dynamics of cells or of the system of vessels, etc.). But observation levels are not limited to the positions we assume in space, because they include all possible conceptual profiles we select. Thus, a commercial manufacturer of plastic flowers will probably select an essential performance in terms of pure appearance, while a manufacturer of educational tools will concentrate on the structure, usually only macroscopic, of the main anatomical parts of the rose. A publisher, in turn, for illustrative aims, will have an expert create an image of some drawing that outlines parts of the rose for the printed page, therefore in two dimensions, while a painter or a sculptor will make aesthetic decisions based on a much free range of interpretations.

It is interesting to note that in children's drawings the essential performances are simple and preponderant, though often one needs an explanation by the child—as is the case with abstract artists—in order to understand the graphic language and, therefore, the theme of the drawing. The fact is that, during the adolescence, the ability to simplify reality first begins to develop. It will later give rise to the selective attitude we are speaking about, useful not only in making naturoids but also in operating in the world of science and technology and, in the end, in surviving in everyday reality itself.

Part III

Chapter 8
The Mind Reduces Complexity, Reality Does Not Make Discounts

In 1983, in one of the rare works dedicated to the artificial, the Italian biologist, M. Rizzotti, though orienting his discussion towards an understanding of this concept, which is rather traditional (as anything made by man), insightfully notes that:

> … whatever intervention [action, technological construction, etc.; Editorial Note] involves, for its own nature, an amount of mass and energy which concerns not only the resulting object but also the environment … If we displace some stones we displace always some amount of mould, we crush some insect, and we lower the soil … Even if these effects are microscopic and secondary, they appear always along with our action (Rizzotti 1984)

This emphasis allows us to introduce, after our having become familiar with the three main concepts of the theory of naturoids—namely the observation level, the exemplar and the essential performance—a principle which is fundamental to our discussion, which we could name the *principle of inheritance*. This expression refers to a circumstance that is very simple in itself but often neglected in theory and practice. Whatever action we perform, including the development of a naturoid, it is not limited to the effects which are predicted and planned by the design. In fact, any action includes many effects, regarding quality and quantity, which come, for inheritance, from all the interactions we trigger at all the known and unknown levels. However, these effects could be predictable *a priori* if, and only if, we could take into account all the involved levels and their interactions, and this is impossible in principle. For instance, in the field of artificial organs,

> A great deal of thought must be given to all aspects of device use, from charging of the batteries to dealing with unexpected events. All these issues must be resolved in the conceptual design phase … (Galletti and Colton 2006)

If you read the description of any drug, you will immediately see what we are talking about. *Side effects* are a very important part of the work of a pharmacologist, who, having to pursue some objective, knows well only some of the effects that his drug will generate. Usually, the range of side effects will be enlarged only by the extensive and lasting adoption of the drug, though it can never be said to be complete.

M. Negrotti, *The Reality of the Artificial*, Studies in Applied Philosophy, Epistemology and Rational Ethics 4, DOI: 10.1007/978-3-642-29679-6_8,
© Springer-Verlag Berlin Heidelberg 2012

Side effects frequently manifest themselves in the form of unpredictable *sudden events*, caused by particular combinations of events which are sometimes explainable but often unexplainable. Such events are common not only in the area of pharmacy and medicine but also in meteorology, engineering and many other fields.

The main point, as far as side effects related to technology are concerned, is that technologists—but this is also true for just about anyone—cannot take actions which, having to achieve some objective, are able to achieve that objective and *only that* objective.

In other words, any action we take, does not just generate the effects it was intended to generate but a lot of other things at an unpredictable number of observation levels. For instance, when someone buys a car for the first time, the car will trigger a series of new events which go beyond his planned aims. As we know, the use of a car changes our habits, provides us with new sensations, can affect our physical appearance, implies a revision of our budget, absorbs time for its maintenance or repair, contributes to pollution problems, and so on. It is not simply possible to buy a car and nothing else. Its 'side effects', those which are already known and those which will come as a surprise, cannot be eliminated.

All this applies to a very wide spectrum of human activities. This is why, for instance, music may have no meaning in a strictly semantic sense (Sloboda 1985), but it surely has extra-musical effects. For example, just as a drug may produce biological dynamics which are different from the action level at which it was designed and tested, music may have an effect on some classes of physiological phenomena.

We could say that reality, as it were, does not give a discount and that our actions always have numerous consequences in very heterogeneous areas, due to the several interactions among the matter involved. These consequences will involve several observation levels which are generally not considered, nor can be considered in their complete wholeness, when we plan our actions.

In the case of a naturoid, the inheritance principle is made possible by the subsequent choices that the designer will inexorably make (selecting an observation level, isolating an exemplar, privileging some essential performance) and also, perhaps above all, by the materials and procedures he decides to adopt. The role of multiple selections and the inheritance principle in giving the naturoid its own properties is well illustrated by the following anecdote, by the art psychologist Rudolf Arnheim:

> The smoke detector in the new library where Mary works proved to be so sensitive in the beginning, that on two separate occasions, when an employee lit a cigarette in the office, firemen were called. Some sensory devices artificially created by man, respond to danger signs with greater reliability than the senses we are born with (Arnheim 1971)

It is clear that no human would have acted in such an exaggerated way, but an artificial device, lacking any discriminatory capacity—without the ability to develop it through evolution—and privileging the essential performance for which it was designed and only that function, is inevitably vulnerable to such reactions.

The same is true of so-called 'errors', which often characterize computers, burglar alarms and many other control devices. In the mentioned cases, in order to get from the device the ability to respond effectively to a given class of inputs and generate the desired outputs, one has to design a structure—that is to say, a set of components, that interact. But the interactions that are mobilized inside and outside a structure do not limit themselves to the ones that were designed. As a matter of fact, the more dynamically complex a machine, the more it generates interactions which are very difficult to establish during the design phase. This is why, by the way, the project for the Strategic Defense Initiative (SDI) was criticized, for it was not possible to test the software except in a real, undesirable situation.

The artificial exhibits its greatest bent to the inheritance principle when the materials are concerned. This issue may be summed up in very simple terms. On the onc hand, we have no a priori opportunity to realize how many observation levels constitute natural reality or, in any case, which ones we should adopt in observing it. Such a list would end up being arbitrary, but, above all, definitely incomplete. For example, how many levels define a stone?

The list could begin at a strictly petrographic level and then move on to the physical macroscopic and chemical levels, including the geological level, proceeding then to the physical microscopic, atomic, electronic, and so on.

In addition, we should consider also the physical environment within which the stone appears in nature, because it could be strategic in order to have it working well in the context we have in mind. This is a well-known deal and very delicate issue when, instead of considering rocks or stones, we have to do with the problem of setting up the right environment for animals within a zoo, which could be placed in a region very different from the one in which they live in nature.

Thus, when we use a stone in a given technological project (in the field of naturoids or in a conventional technological field) we do so because some of its properties appear suitable for some aspect of the design itself at an observation level which is suitable to our needs.

The properties that draw our attention, however, do not exhaust the 'nature' of the stone in question. The properties which would attract the attention of the technologist, in fact, will be perceivable at some observation level, but, if we shifted to another level, then we would discover dozens of different properties of the same stone. Some of these hidden properties could, *a posteriori*, make the achievement of the planned aim of the design easier; others, on the contrary, could prove to be obstacles, while others might be neutral.

Inevitably, when we adopt the chosen material, for building an artificial stone, we will *inherit* all its properties and not only the ones which attracted our attention. The inherited properties may remain dormant and silent for an unpredictable amount of time, revealing their presence and affirming their 'rights' in special circumstances which are quite unpredictable. The impossibility of their *a priori* description depends, as it is easy to argue, on the number of possible interactions which the various levels of the stone could have on each other and, furthermore, on the levels of the real environment the artifact is hosted in.

In the end, the number of the possible interactions cannot be calculated, because, if we assume that any natural exemplar is characterized by infinite possible observation levels, then its encounter with any real object else—which is in turn characterized by potentially an infinite number of levels—will produce a quantity of possible interactions which is equal to the product of two infinite numbers, which is an undetermined number.

In the field of construction materials, it is well known that some marbles are more sensitive to rain and to its chemical components than others, which are more sensitive to other organic or inorganic natural phenomena. In all cases, the result is that, after a period, the appearance and sometimes the structure itself of the building is strongly modified by these undesired reactions. Mechanical or chemical procedures adopted for cleaning such buildings, in turn, lead to new problems. Thus, as has been observed in some studies carried out by the Masonry Conservation Research Group of the Scottish Robert Gordon University, in some circumstances buildings were damaged to such an extent that the stones rapidly decayed. Indeed, according to the study, cleaning work was carried out quite ignoring its effects and the consequences.

Even artificial marble, which dates back to the Baroque period in Europe, needs to be cured for, because of

> … degradation and weathering of artificial marble by environmental and climatic elements. Water and salts from rain and rising humidity from the ground, by deposition of aerosol and gases, as well as changes in humidity and temperature are responsible for weathering of building stones and plasters. The main reasons are hydric and thermal swelling and shrinking processes as well as crystallization of salts (Wittenburg 2000)

Current plastic marble imitations certainly would have their own story to tell in this regard.

Let us consider some other examples. A contemporary operating room, unlike those of past centuries, is a very controlled environment because today we know that whatever object or surgical instrument is used by the surgeon, it does not just possess the desired properties (for instance, mechanical). It may also have other rather dangerous properties, including microbiological entities, dust, metallic residues. However, despite all the precautions which are taken, no operating room in the world may be said to be completely controlled, since, by definition, we can only control the phenomena we know and, sometimes, only partially.

Even here, in other words, we have to admit that the most reliable way for getting some guarantee of the compatibility of our actions with our desires and with the environmental requirements is only to be found through extensive and lasting experience in the field.

The most evident phenomenon is in a medical or biological field the various kinds of so-called 'rejection', including reactions of the immune system. Rejection is a true rebellion of the organism, not against the essential performance that, for instance, an artificial heart may exhibit, but against some component of it which is perceived as extraneous or dangerous to the body. Tony Keaveny, from the University of California, Berkeley, sheds light on the issue in describing the problems that bioengineers face in their attempts to build and to place artificial bones into the human organism:

> … joints are trouble-free for 15 years [and this may be evaluated as] a remarkable record considering the harsh biomechanical and biochemical environments of the body (Keaveny 1996)

Furthermore, are multi-level interactions always around the corner because

> Failure of artificial joints is a multifactorial process involving a cascade of biochemical events (Prendergast 2001)

Because of the immune system reactions, other researchers at Rice University Institute of Biosciences and Bioengineering remember that until recently:

> … most research in the field [of cell transplantation] has focused on minimizing biological fluid and tissue interactions with biomaterials in an effort to prevent fibrous encapsulation from foreign-body reaction or clotting in blood that has contact with artificial devices. In short, most biomaterials research has focused on making the material invisible to the body (Mikos et al. 1996)

A biomaterial was defined by the American National Institute of Health in 1982 as

> … any substance (other than a drug) or combination of substances, synthetic or natural in origin, which can be used for any period of time, as a whole or as a part of a system which treats, augments, or replaces any tissue, organ, or function of the body (NIH 1982)

The same institute underlines that

> Materials science was defined as the science which relates structure to function of materials … The field of biomaterials is first and foremost a materials science. (ibidem)

and that

> In evaluating safety and effectiveness of biomaterials, the material cannot be divorced from the device … Each biomaterial considered for potential clinical application has unique chemical, physical, and mechanical properties. In addition, the surface and bulk properties may differ, yielding variations in host response and material response (ibidem)

Actually, at a microbiological observation level, interactions are always at work, and, for instance, interactions between biomaterials and neutrophils includes oxidative bursts, degranulations, neutrophil functional changes, and changes in neutrophil lifespans (Videm 2008).

It is important to note that, though the statements by Keaveny and Mikos were made in the second half of the 1990s, bioengineers are currently experimenting with new strategies for getting biomaterials which may be called 'hybrid'. This means that the new materials aim to harmonize the artificial organ and the host environment, making them compatible.

Using suitable biomaterials, they try to give, for instance an artificial cell, sufficient compatibility with the host organism at the surface level, while maintaining the needed artificiality in its internal structure. This is another example of an attempt to deceive nature, since the organism will be induced to accept the performances generated by the naturoid cell without attacking it, because its interactions with the cell will be mediated by a compatible surface. 'Hiding' the device is another expedient.

> … insulin-secreting cells form pancreatic islets, usually taken from a pig, in semi-porous capsules implanted in the body. The capsules must be biologically and chemically inert; that is, their chemical composition cannot induce inflammation or other reaction from the body, and they must resist decomposition. The capsules must contain pores small enough to exclude the mobile cells of the immune system, macrophages and lymphocytes, but large enough to allow a physiological release of insulin in response to blood glucose levels (Edwards 2001)

The possible extension of such a hybrid strategy to other areas of naturoids raises interesting possibilities, though not completely new ones, if one considers, among the many cases we could cite, the attempts to make artificial intelligence programs or robot something friendly or even anthropomorphic (that is to say, externally similar to human beings by some kind of interface and even with some special skin-like covers). In all cases in which designers resort to such interfacing strategies—between the naturoid and the natural world—their meaning should in fact be found in the effort to make compatible two heterogeneous realities, at a macroscopic observation level, whose interactions are controllable only within a very narrow range.

Regarding the functioning of artificial cells, McGill University researchers (where Thomas Chang began this kind of research in 1957), claim that artificial cell membranes can be significantly modified by adopting biological or synthetic materials. Their permeability can be controlled in many ways. In this way, the materials enclosed in the cell can be held back and kept separate from undesired external materials. On the other hand, Chang himself said in 1996, in contrast with de Monantheuil's statement cited at the beginning of this book,

> … if you look at it objectively, no matter how smart we are, we will never be able to copy what has been made by God, not even a simple red blood cell. We can only hope to make a simple substitute, and right now we are still taking our first steps (Chang 1996)

In any case, it is clear that every concrete attempt to reproduce a natural exemplar, and a performance of it which is considered as essential, implies the generation of a set of realities much richer than one might desire.

We can also interpret along the same lines the attempts, which appear very frequently throughout the history of technology, to build automata—almost always 'androids' built for entertainment objectives—just consider the first android built in the sixteenth century by H. Bullmann of Nuremberg, the automata by Gianello Tornano of Cremona or the copyists of the Swiss Pierre Jaquet-Droz, the flute player by Vaucanson, and many other examples from the eighteenth century (Bedini 1964; Losano 1990). These attempts that could be understood as true, friendly 'interfaces' of mechanics with humans, furthermore, converge with all the tradition that goes back to the mannequins universally used in many kinds of theatrical performance and the like, or rites, or, today, for emulating situations and events in medicine, in car safety testing or in military training. Though the issue is somewhat extraneous to our discussion, it would be very interesting to ask ourselves what relationship might exist between all this and the very ancient inclination of man to build camouflage devices which, on the one hand, imitate the human face but, on the other, transfigure his own identity, aspect or presence. It is a matter of very diffused habits dating back to the most ancient masks used in religious ceremonies in China, in funerary Egyptian art or in the Greek theater, and, in general, in all the world's cultures at some point in their development.

Chapter 9
The Problem of Synthesis

The situations within which scientists and technologists act, above all those regarding the technology of naturoids, are surely complicated, permanently, by our incapacity to place ourselves simultaneously at more than one observation level. Though scientific methodology has developed several techniques for controlling more than one variable, *de facto* our theories and models always unavoidably focus on some aspect or profile that is always considered as central. All you have to do is consider the history of recent science to realize how true is.

The fact is that attributing some privileged and comprehensive meaning to the events we observe at a given level is a limiting process which prevents us each time from looking at these events from other levels, if not at different moments, therefore raising the problem of their logical, functional or operational connection.

The tendency to assign some sort of privilege to our selections of observation levels is a 'temptation' that has appeared many times in the history of scientific ideas: for instance, in the Nineteenth Century, which attributed a central role in a number of phenomenologics to electricity (Leschiutta and Leschiutta 1993).

Today, the dominant theme, which we have already referred to, appears to be information, to which is reduced a series of phenomena such as intelligence and communication, the intimate stuff of social relations and often even art. Stating it very roughly, the Eighteenth Century was largely dominated by a mechanical observation level, whereas the nineteenth and twentieth were dominated by electrical and information levels, respectively.

The history of the damage or, at the very least, the waste in terms of research projects caused by these reductions or polarizations, has yet to be written, but, in the end, the human inclination that produced them—that of privileging one observation level at a time—definitely seems to be without alternatives.

If we consider the world of naturoids, the above-mentioned inclination goes hand in hand, as we have seen, with another unavoidable constraint: the one which forces the artificialist to use different materials and procedures than the natural exemplar and, thus, to introduce to the object or process which is to be built, a

M. Negrotti, *The Reality of the Artificial*, Studies in Applied Philosophy, Epistemology and Rational Ethics 4, DOI: 10.1007/978-3-642-29679-6_9,
© Springer-Verlag Berlin Heidelberg 2012

tendency to generate side effects and sudden events whose frequency, intensity and quality are unpredictable.

In the best case scenario, it is only possible to limit these effects, as we have seen in the previous section, by means of a sort of 'encapsulation', i.e., isolation of the artificial from the external world, for instance the host organism. In such a way, the only interaction which can take place—between the artificial and external world—is the one we defined as essential performance, the only umbilical chord connecting, to some extent, the artificial and the natural. This solution, on the other hand, will place a naturoid in a very unnatural position, since within nature no phenomenon can be completely isolated from the others because the relationships among the several observation levels involved are, so to speak, always at work.

All this leads us to another question. In general terms, under what conditions and with what results, could one design more exemplars and more essential performances which cooperate in a unique artificial system that reproduces some natural system?

If we take the case of a flower which we want to reproduce at more than one observation level by means of materials and procedures different from the natural ones—for whatever aim, scientific, educational, commercial, etc.—then we would face a fundamental difficulty. For instance, we must decide what kind of relationship among its parts (stamen and carpels, style, ovary, stigma) we should reproduce. In other words, at which observation level (cellular, molecular, etc.) should we reproduce the relationships that characterize the flower as whole, in order to accurately reproduce not only its structure but also its functions?

Clearly, presented in this way, the problem is theoretically and practically irresolvable, since, if an artificial flower must be a reliable anatomical and physiological reproduction of the exemplar, then, if we succeeded in this sort of reproduction, we could claim that we have replicated it, i.e., recreated it entirely. The analytical rebuilding, piece by piece, of a living system—but also any other sufficiently complex natural reality—starting from its basic chemical elements, is a task which is definitely beyond our capacity and, perhaps, intrinsically impossible. Indeed, genetic engineering may be able to achieve this kind of reproduction, but, of course, in this case, we do not rely on different materials or analytical models of the various sub-systems the whole system should contain. In genetic engineering we only set up the conditions through which nature works by itself.

But the designers of naturoids have a very different aim. In principle, what they are attempting is the replacement of natural materials of the exemplar with other natural materials which approximate them at a given observation level. Our current knowledge of organic and inorganic materials and our ability to manipulate their features—enhanced a great deal, by the way, by space technologies—allows us to generate substances and physical or chemical structures, in many fields, which are very similar to the natural ones. However, this similarity is almost always recognizable at the selected observation level.

Returning to our case of a flower, today we are able to generate a wide range of artificial scents and, among these, we could select the one which is the most suitable for reproducing our natural exemplar. But, if our aim is to rebuild even only the structural and physiological sub-system which produces that particular scent in the flower, then we are faced with quite a different problem. We must decide what relationship to establish between the artificial scent and the artificial structures we would have set up in order to allow those structures to generate our artificial perfume, rather than simply pouring it on them.

At this point, it is clear that the design of the partial artificial objects (scent on the one hand and some anatomical structure on the other) would be contorted, since their reciprocal dependence in the new possible design will surely impose an indefinite quantity of adjustments.

A huge set of new problems would then arise, and, these problems, would require some drastic decisions. For example, we might decide to establish an anatomical threshold under which we might give up the criterion of similarity with the exemplar, thus limiting the reproduction to some more superficial aspect suitably described by a reasonably simplified model.

As in the artificial flower, some artificial device would be made to generate the scent while remaining invisible within the whole organism. Even if its characteristics were compatible with the system, other artificial parts might be forced, through some expedient, to come into contact with the scent in a way sufficiently similar to the natural way, at least in terms of the new observation level which we would have generated or selected.

It is important to note that the observation levels at which the two partial artificial objects are reproduced will no longer have any relevance in the subsystem we have built. They will be 'absorbed' by a third level: the level at which the relationship between scent and anatomic-physiological structures of the flower become possible. The two original exemplars and related observation levels will be, as it were, sacrificed for this relationship. They will be remodeled in order to serve the new essential performance, which is constituted exactly by the relationship between the two exemplars.

As we have seen in the unavoidable complications of the above-mentioned methodological procedure, the cooperation between two artificial objects or processes poses serious problems. Man, almost invariably, tries to overcome these obstacles by resorting to a decision which establishes some definite objective to reach.

On the one hand, this strategy constitutes a *de facto* renunciation of the reproduction of the exemplar in its entirety and, therefore, its replacement with a simplified model which privileges only one observation level. On the other hand, such a strategy implies the tacit admission that, if one wishes to integrate even only two observation levels, one must proceed, when it makes sense, to establish a third level, without assuming that it fully incorporates the previous two. Being able to completely rebuild an exemplar using a bottom-up strategy is, to sum up, a pure utopia.

Let us imagine designing the reproduction of the sub-system which controls and coordinates vision and touch in the human body, provided we know it well enough. Obviously, all this will require setting up a model in which the relationship in question is central, assuming that the artificial devices for vision and touch are reliable enough. Such a model is quite plausible, thanks to the availability of electronic and computer technologies able to implement many types of very flexible complex algorithms. But the critical point is something else. It would have to deal with the necessary adjustments we make in the artificial vision and artificial touch devices in order to make them compatible with a third artificial device, which adopts some region of our brain as exemplar.

In short, the true exemplar of such a design, would be the brain region itself and this will not lead to a pure 'sum' of the two partial artificial objects already available, which were designed and built as separate devices, namely the artificial vision and touch devices.

The coordination device, in such designs, is usually electronic or computer based, because, as we have already mentioned, this gives researchers today the most powerful working tools.

But the electronic or informational levels, which are certainly present and relevant in brain activity, do not exhaust its reality, since the brain works at other levels as well. Furthermore, the electronic and informational levels in question will introduce, by inheritance, their own natures, which will also be imposed on the two partial artificial devices (for vision and touch) that, in fact, will be coordinated through some electronic circuit and computer program.

From this point onwards, the typical series of alterations of the sub-system, as compared to the natural one, will begin.

Thus, to cite a real case, the anthropomorphic robot Hadaly (built at Waseda University, Tokyo, in 1995) is structured in three sub-systems: a vision sub-system, an auditory sub-system, and a motor. But, beyond some minimal supervising algorithm of its behavior, Hadaly does not reproduce at all the performances we would expect from a human being, with his capacity to coordinate, both on a reactive and a high-level decisional basis, the events belonging to the three levels involved.

If one asks Hadaly where some department of the university is located, it tells him and even points in the right direction with its hand, but if one approaches its hand with a lighter, Hadaly does not draw back or protest. Clearly, the introduction of an additional program in its computer, able to evaluate this kind of situation, would not solve the problem at all.

In scientific terms, i.e., looking at nature in a way which aims not to reproduce it but to know how things are, the logical aspect of the problem we are discussing is in some measure the same. As a typical case, let us cite the problem posed by a biologist referring to the relationships between vision and touch in the nautilus:

> Another fascinating problem is the relationship between visual and tactile learning … Since the two systems overlap in the vertical lobe, maybe there is some kind of coordination between them. However, it has been demonstrated that the objects detected by sight are not recognized by touch (Young 1974)

However, even here, scientific problems are paralleled by concrete questions related to the design of naturoids, because the simultaneous putting into action of two or more artificial devices poses the significant problem of dealing with their *real* interactions and, therefore, are not merely hypothesized. In general terms, it has long been observed that such a:

> ... multi-purpose approach is quite different from the single-function concept that governs most machines. Designers must always be mindful that each organ works in conjunction with all others: liver failure damages the brain and kidneys; the lungs and heart function in an intimate way; etc. (Fitzgibbons 1994)

In the second part of this book we will examine several examples of artificial devices within which one may guess many of the above problems and strategies are to be found.

Let us make some further observations regarding the possibility of obtaining cooperation from two or more artificial objects or processes in the same natural organism, for instance the human body. In principle, it seems that one would have little difficulty in linking two artificial objects, say A1 and A2, which had proven to be effective as standalone devices, to work together. In fact, 'implantation'—to be distinguished from 'transplantation', which implies placing a natural organ in a body—of an artificial bone and an artificial heart in the same organism should not give rise to any problems. However, the implantation of an artificial duodenum and an artificial liver—though they are only available as experimental devices—would be a very different case, full of snares and degenerative possibilities due to the functional relationship that exists between these two organs in the human organism. We can surmise that the cooperation between an artificial liver, duodenum and pancreas would be even more complex.

Once again, the general problem is the variety of the observation levels involved and the arbitrariness that will govern the choice of the essential performances. Any model was used to pilot the design would only be able to select a basic level, relegating the others to the background. The resulting artificial device—a true artificial sub-system—would work well only until the assumed exemplar and its related essential performances, were determinant for the functional balance of the organism. Nevertheless, if some different performance were needed—an essential performance which would have required a different, more complex model, the sub-system would start a process of uncontrollable degeneration of itself, the host organism or both.

In conclusion, in an artificial sub-system consisting of more than one partial (local and separate) artificial device, the greater the *functional distance* among the organs chosen as exemplars, i.e., the more independent they are of one another, the greater the likelihood that the sub-system will function properly. But, this is clearly a rather ambiguous and uncertain condition if one considers the deep interconnections which characterize natural reality in all its phenomenologies.

Chapter 10
Transfiguration

A famous Latin saying states: *senatores boni viri, senatus mala bestia* (senators are good men, but the Senate is a bad beast). In general terms, this means that the coexistence of single entities of a given kind may give rise to a very different sort of whole which cannot be explained by or limited to, the 'qualities' of its components considered individually.

Usually this phenomenon is defined by the term *emergence*, coined by G.H. Lewes in 1875. Often it is a matter of a simple change of observation levels. For instance, a mass of white and black microscopic granules will appear as such under a microscope, but, to the naked eye, the mass will appear gray. Nevertheless, in other cases, it cannot be denied that, though one remains at the same observation level, the 'sum' or the 'synthesis' of many objects or processes gives rise to something which goes beyond the features of the single parts. After all, chemical compounds well describe this fact.

However, the principle of emergence constitutes a general foundation on which the hopes of designers of naturoids in several fields are based. For instance, in 1987, Craig Reynolds demonstrated that the coordinated flight of a flock of birds can be simulated without including any central coordinator in the model. Each simulated bird (or 'boid', as Reynolds called them), follows the following simple rules: avoid collision with other birds (collision avoidance), stay in step with other birds (velocity matching), try to stay as close as possible to other birds (flock centering). Lastly, each bird could only see its closest friend. The simulation—which was then adopted for successful movies—demonstrated that a flock organized in this way was able to 'fly' on a computer monitor as a compact whole, with birds successfully avoiding obstacles to rejoin the flock just after them, exactly as it happens in observable reality.

Other models—always formal, i.e., simulated on a computer—in the research area which is defined as *Artificial Life*, or, in short, ALife, are able to reproduce typical phenomena of living processes (self-reproduction, evolution, the struggle for survival, etc.) as they emerge from the coexistence of many single 'agents' (cells or 'cellular automata') and, therefore, from the relationships which arise among them.

M. Negrotti, *The Reality of the Artificial*, Studies in Applied Philosophy, Epistemology and Rational Ethics 4, DOI: 10.1007/978-3-642-29679-6_10,
© Springer-Verlag Berlin Heidelberg 2012

Chris Langton defined ALife as:

> … the field of research dedicated to the understanding of life through the attempt of abstracting the basic dynamic principles which stay at the basis of the biological phenomena, in order to recreate these dynamics in different supports—like computers—making them accessible to new manipulations and experimental tests (Langton 1989)

The spontaneous emergence, as a self-organizing phenomenon of intelligent behaviors or behaviors typical of life, sometimes depends on the achievement of a sort of 'critical mass'. This is what occurs in crowds or in nuclear chain reactions, or in all complex phenomena, in the evolutionary differentiation of living systems and in the emergence of intelligence in the human or animal brain.

Though the 'new types of manipulation and experimental tests', which Langton refers to, are virtually non-existent—if by 'experimental' we mean real tests carried out in concrete terms, and not just informational tests—the ALife researches provide us with further food for reflection in the field of naturoids.

In addition to the hopes that it generates in the field of artificial intelligence or ALife research—regarding the possibility that intelligence or life might suddenly and spontaneously emerge from their models—the principle of emergence may be useful for emphasizing a very simple truth, but, even in this case, permanently underestimated. We refer to the fact that in the field of concrete phenomena, and, in some ways—as we saw with ALife though for other reasons—in any system we build, it invariably ends up by giving rise to something which goes beyond the objectives of our original design.

Self-organization is also central in the so-called 'genetic algorithms'. They are computer programs which, through a set of simple rules reproducing the rules of Darwinian evolution, are able to find solutions for complex problems difficult to solve through ordinary analysis. Through progressive processes of mating, crossover, and selection of best fitting cases, a population of members adjusts its behavior to reach the target. Such a solution is said to emerge from the behavior of a critical mass of members acting under the rules of evolution.

The hope of designers of naturoids, mainly in the software area, is sometimes that, given the right mass dimension, and of course, the right model and the right rules, some performance of the exemplar could appear in the final naturoid, even if it has been not intentionally designed.

In this sense, emergence may be understood as an extension, or a particular case, of the inheritance principle. Actually, the hope of designers is that the inheritance of the properties could autonomously orient itself towards some undesigned, but meaningful outcome.

In other words, at a given observation level, the features of a system constituted of a certain number of components may appear new compared to the features of the components as such because the latter properties belong to another observation level. This happens because the relationships among the components may exhibit their own qualities, starting from the relationship itself, independently of the characters of the components in question.

Experience, even everyday experience, continuously shows us how true this is. Just consider the circumstances we might encounter in a laboratory or even a kitchen in which, on the basis of chemical or physical knowledge, we face reactions which produce new realities quite different from the simple 'ingredients' we have used. Likewise, in the sociological field, we know well how certain collective phenomena—panic, aggression, fanaticism, etc.—are due to psychosocial relationships which have no correspondence to individual motivations or attitudes: collective phenomena, clearly require relationships between individuals, and this involves a shift of observation level.

Of course this is even more true in technological areas: the *blackout* which paralyzed New York in the 1960s, as well as the many sudden break-downs which strike machines or systems of many different kinds, generally belong to the same typology, i.e., to a class of events which emerges from complexity and follows its own logic, unpredictable and therefore uncontrolled by the single components. A naturoid is really nothing but a particular case of this typology, of course.

The point is that the quality of what may 'emerge' from an artificial device—i.e., the additional performances it may exhibit as compared to the essential performance included in the design—will simply not just consist miraculously of something similar to what the exemplar exhibits in nature.

In fact, what happens in genetic algorithms and ALife programs is made possible by the intelligence—due to Darwin's theory—of the rules governing a process which consists of information driven by those rules. In other words, a computer program is a reality characterized by only one observation level—namely information—and, therefore, the rules are able to drive the events according to their own nature without any interference coming from other observation levels. In the real world, there are no standalone dimensions or levels, and therefore, no algorithm can work undisturbed. This is why, by the way, natural evolution takes a lot of time and several species fail to survive. All that can be said is that in the real world 'something always happens' right, because of the extremely dense set of multi-level interactions that characterizes nature.

The accurate reproduction of two natural organs—at their own observation levels with their own essential performances—which, in the human organism, also work together to regulate some other process, does not at all guarantee that the regulation process or performance will emerge automatically from their coexistence. This will occur if, and only if, a precise condition is satisfied: the expected performance must be a simple function (or consequence) of the essential performances reproduced by the two artificial organs.

To cite a rather general example, if we accurately reproduce some performance of sunlight (e.g., arrays belonging to a well-defined region of the spectrum which stretches from ultraviolet to infrared) and, simultaneously, a given quantity and quality of heat (dry, moist, airy, etc.), it is possible to obtain an 'artificial climate' suitable for the growth of a certain tree.

Indeed, we could state that an artificial climate will emerge from the combination of the two artificial processes in question, but it will be an emergence, as it were, guided by a design founded on sufficient knowledge to generate that phenomenon and only that phenomenon. Actually, as far as artificial light is concerned, we have to accept the idea that it is always different from sunlight, and thus other things will happen, or emerge, because:

> … in the absence of bad weather, nocturnally migrating birds have been observed to be confused by artificial lights below them and are attracted at night to artificial lights when there is no moon combined with fog or mist at ground level in the area. [In short, birds are] … drawn to light due to the differences in the properties of natural vs. artificial light (Verheijen 1958)

Furthermore, it has been noted that artificial light radiations do not generate a complete spectrum since they favor some colors at the expense of others—orange, yellow and red in the case of incandescent light and yellow–green in the case of standard fluorescent light. All this may surely imply something new in adopting artificial light in a variety of situations, but what will emerge only by chance can be significant for a system.

In short, all science consists of a description of phenomena of this kind, though, very often, we are not able to give some analytical explanation. This is what happens with many drugs, therapies and physical products or processes we are able to produce by suitably combining some objects or events though ignoring their genesis beyond the observation level at which we orient ourselves in order to obtain them. From among the more spectacular cases, just consider aspirin. We continue to discover more and more new effects of aspirin, including the recently discovered unimaginable reduction of the incidence of heart attacks in diabetic patients.

In our opinion, the most interesting aspect of the matter we are discussing here is the fact that we only know about researchers' success and do not consider their failures, which are certainly much more common. More precisely, what has 'emerged' from scientific or technical attempts, carried out throughout many centuries of research, that have failed in the sense that they have not led to any interesting or useful phenomena? This question is very important if we consider how often it happens that experimental projects with specific objectives lead to 'emerging' phenomena of an altogether different kind which are considered to be so interesting that research shifts its focus to these phenomena, as in the chance discovery of the semi-conductivity of doped silicon. In these cases, research fails in terms of fulfilling its original purpose, generating discoveries that lead to research regarding other purposes which do however produce very significant knowledge.

Even in all the other cases of failure—those in which the failure is not accompanied by unexpected discoveries—*something certainly does happen*. This is also true for the design of naturoids: the failure of a design in this field does not mean that 'nothing happened', but that the performance which emerged had nothing to do with the performance considered to be essential in the natural exemplar, or was only partially or weakly related to that function.

The principle of inheritance, on the other hand, draws our attention to the fact that, even in the case where we are able to satisfactorily reproduce the selected essential performance, an unpredictable series of other performances is, so to speak, lying in ambush when it is not present in an apparent way. This in turn means that the interactions between the naturoid and natural world, including therefore interactions with human beings, will always depend on a much wider spectrum of performances than expected in the original project. In this framework, a naturoid which does not reproduce the selected essential performance of its exemplar is certainly a failure, but, after all, it is only a particular case of all the artificial objects and processes.

Naturally, this implies the possibility that, among the performances which come from an artificial object, some may be acknowledged as typical of the exemplar, even though they have not been designed intentionally. This possibility is obviously unthinkable and unpredictable. It can happen, but not thanks to the precision of the artificial objects and processes already reproduced, since, as we have emphasized many times, it is very rare that a given performance is the simple, additive function of two other performances. On the contrary, it is more common that the performance in question, instead, is located at a different observation level than the ones adopted to design the two naturoids and the related essential performances. The essential performances of the two naturoids, in other words, can sometimes be adopted as the necessary conditions, which, however, are not enough for the emergence of a third performance similar to the natural one.

However, what is certain is that both for the principle of inheritance as such and for the principle of emergence—intended as an ineliminable derivative of any recombination of natural things—any artificial object or process can only be intrinsically intended to generate a *transfiguration* of the exemplar and its performances. Moreover, the transfiguration is even more amplified by the inevitable use of conventional technologies. As we have seen from the beginning of this work, conventional technologies can clearly be distinguished by their determined heterogeneity regarding the structure and dynamics of natural objects and processes which, in fact, such technologies intend to control and modify rather than imitate or reproduce. It is worth noting that the transfiguration which we are referring to is not, in itself, negative. Very often the technology of naturoids produces objects, processes or machines which exceed the performances of natural exemplars which they arise. This is the case for computers, machines which, among other things, reproduce the essential performance of logical or mathematical calculation, with a speed and precision clearly superior to man's. And this is also the case for more concrete artificial objects, such as those which are inspired by the certain abilities of animals or of humans, such as taste or smell, vision, and so on.

However, transfiguration can only be avoided if, in selecting an observation level, we actually are able to isolate our exemplar, and its essential performance, from all the other observation levels which characterize natural reality, thus defining a rather new, purified and stand-alone feature. This is, obviously, impossible working in the real world, while apparently it is easy when we simulate something on a computer, or when, in scientific theories, we resort to abstractions substituting the world with a simplified modeling of it. These are extremely useful undertakings and procedures which, in any case, are the only alternatives to human beings. However, they are also procedures which should not be confused *ipso facto* with discoveries of nature, even though at times they allow us to control it in well-defined circumstances.

Thus, in the case of Reynolds' *boids* and other ALife software creations, we find ourselves, in some way, in the same circumstance as the people who design programs capable of certain logical operations, such as deduction. If we enable the computer to make deductions and we tell it, on the one hand, that "all capital cities are big cities" and, on the other hand, that "Paris is a capital city", it will be able to inform us that "Paris is a big city", even though we never gave it such information directly. The fact is that the emergence of such information is essential to the deductive algorithm which controls the computer, but such an algorithm does not necessarily reproduce the human way of performing deductions. Men, probably, adopt a more biological and complex way than computers.

It is therefore legitimate to affirm that, even regarding computer simulations, although the selection of a single informational level obviously avoids transfigurations on other levels—for the simple reason that other levels, besides information level, do not exist—eventually some form of transfiguration—such as processing speed and precision—still enters into play. This happens because the informational level, although it is fit to describe certain aspects of reality, once it is set up as a unique dimension, ends up itself by living its own life. The capability of a simulation model to generate behaviors which are comparable to the observable natural reality, at this point, is no different from the ability of mathematics to describe the natural world and therefore neither can it be defined as a discovery of the way in which nature functions intrinsically nor is it a reproduction of the world.

For instance, today, according to Michael Tabor of the University of Arizona, Tucson, it is possible to mathematically describe phenomena which are not decipherable in other ways, such as the twining of tendrils which seemingly follows the same rules as those according to which a few spirals of bacterium or even our telephone cables twine themselves helicoidally. But this does not mean that it is able to reproduce a tendril. First of all, as the biologist Neil Mendelson, from the same university, commented, it is necessary

> … to describe exactly what happens to fibers in the real world (Tabor 1999)

We may conclude this section by stating that in no case of concrete reproduction of a natural exemplar, can we avoid some more or less important transfiguration of the exemplar and of its performances. At all steps we have outlined in the preceding pages, selections, or arbitrary attributions, lead in a cascading way to the final design narrowing more and more of its features as compared to those of the natural exemplar. Namely, this selection process involves the choice of the observation level, the isolation of the exemplar, the attribution of an essential performance, the setting up of the model, and, last but not the least, the choice of the materials and building procedures one believes to be the most suitable. This fact will be even amplified if the project of a naturoid will involve two or more exemplars, and therefore, two or more essential performances, because of the methodological and practical difficulty in establishing the right relationships between them. As a result, the area of good functioning of a naturoid—the area within which it can reproduce the performance of the natural exemplar—will shrink and will be surrounded by a progressively growing area of undesigned features.

Part IV

Chapter 11
Classification of Naturoids

Using our reasoning, we can now propose a classification of naturoids on the basis of their main features.

First of all, we have seen how the technology of naturoids has always led to two opposite kinds of activity depending on the *concrete* or *abstract* nature of the 'substance' by means of which the final product is designed and realized. Since man has always possessed and shown a distinct tendency to reproduce whatever surrounds him—and also whatever has a primary origin in himself, such as feelings, self-portrait, etc.—it is not surprising that the entire history of man is intensely characterized by the invention and development of the most varied technologies aimed precisely at the material expression, communication or reproduction of exemplars of every kind.

The development of rock carving and oral communication, the evolution of symbols and painting, the birth of music and poetry, up to the invention of printing and then radio and television, and the advent of informatic and telematic systems, are a few of the cornerstones which have marked man's effort to communicate, that is to say to 'make common' subjective and objective exemplars and essential performances.

In spite of their incorporation into material objects and processes—rock or paper, sound, voice or colors, electromagnetic waves or electric signals—all forms of the communication or simulation of reality can be considered as abstract artificial objects and processes for the simple reason that their purpose is not to reproduce the exemplar concretely and materially, but to reproduce our representation of it *as such*.

For example, when we decide to communicate the images we see and feel sensations while admiring a certain flower, we obviously do not have any intention or ambition of reproducing that exemplar materially. In this case, each one of us—but also the poet or the painter, the writer or the computer simulation expert—only tries to share his own portrait of the world, either by objective pretenses ("now I will tell you how matters stand") or by openly subjective intentions ("now I will

M. Negrotti, *The Reality of the Artificial*, Studies in Applied Philosophy, Epistemology and Rational Ethics 4, DOI: 10.1007/978-3-642-29679-6_11,
© Springer-Verlag Berlin Heidelberg 2012

tell you how I see the world"). While scientific models, including even the most varied 'theories' belong to the first class, the artist's 'poetics', as well as each of our personal expressive styles, belong to the second.

A very different situation arises when someone decides to attempt the material, concrete reproduction of a natural exemplar. In this case as well, as we have already emphasized, we inevitably use representations (in the form of models) and, therefore, we introduce a subjective and arbitrary dimension. Nevertheless, our aim is to build something concrete which hopefully everyone will agree on, so that anyone can recognize, in the object or process that we realize, an object which is part of any common experience: a flower or a heart, skin or rain, rocks or flavors, but also, a little more ambiguously, intelligence or reasoning.

We have also seen, or at least mentioned, how the designer's strategy can point to either an *analytical* reconstruction of the exemplar, or a purely *aesthetic* reproduction.

From the very beginning the aesthetic reproduction privileges appearance—obviously verifiable on a well-defined observation level—of the exemplar, quite neglecting its structure. Examples of this kind are found everywhere: in sculptures, in toys, in architectural 'relief models', in various kinds of prosthesis, in the wide field of so-called reconstructions or 'remakings', various types of gadgets or counterfeits and in many industrial products known as 'imitations'. Usually, the essential performance which stays at the core of this kind of artificial object is clearly commercially, rather than scientifically oriented, i.e., widely perceived in terms of economic demand which, being placed at the same observation level as the producer, grants its success.

In contrast, an analytical artificial object pursues an aim which, apart from its aesthetic appearance, coincides with the exemplar's structure or, at least, with as much of its structure as is necessary to make the required essential performance possible. Therefore here structure does not mean the exemplar's form but, in a broad sense, its anatomy and physiology: thus structure is synonymous with all the correlated parts, appropriately described on an observation level, in terms of material composition and functions.

The matter is not, however, altogether linear and simple. In some cases, the designer of naturoid knows perfectly well that if he does not assign a shape to the artificial object which is similar to the exemplar's, the essential performance will not be achieved. In the case of certain artificial bones, as in many other bioengineering cases, it is obvious that the artificial device cannot have an arbitrary structure compared to the natural exemplar, nor physical dimensions (mass, density, etc.) that make its implantation and functioning impracticable in the human organism. Even for educational aims, bone sub-systems have to respect an obvious essential performance: that of the overall dimension and that of all its parts. Industry products for educational use currently include several bone sub-systems: the artificial skull of a fetus, an artificial human skull, male and female,

> ... constructed from 14 individual parts, which can easily be dismantled and put back together by way of interconnecting plugs (Somso 2001)

as well as an artificial human skeleton, a skeleton of the foot, a skeleton of female pelvis, a hand skeleton and even an unmounted human skeleton.

In the case of the artificial eye—which is not a simple prosthesis but is intended to provide eyesight—it is even clearer how anatomy and physiology place precise structural and functional constraints on the designer, while the reproduction of other organs do not often show any exterior resemblance to the exemplar and are actually intended to be placed outside the organism.

If we then go on to the wide kingdom of natural objects and processes of an environmental kind (rain, snow, grass, landscape, etc.) which do not have a mere aesthetic or spectacular purpose but rather a realistic purpose in the most scientific sense of the word, it becomes clearer that the structural and dynamic knowledge of the exemplars cannot be ignored. All this does not mean that the resulting object or process 'seems' aesthetically similar to the model. Rather, the essential performance of the exemplar and the reproduced essential performance will have to be similar.

In short, isomorphism—that is, the equality or at least the formal analogy of the structures—is indispensable only when function is strictly connected to form and the artificialist intends to reproduce such a performance precisely in the same way. This happens when one attempts to reproduce the fingers of a hand using structures which, though very different from the natural ones, must bear some similarities with real fingers. This is due to the fact that the naturoid must present dimensions which are similar to the exemplar and allow for a few physical performances or freedom degrees typical of natural fingers (prehensility, and opposition of index finger thumb, etc.). In other cases, such as with a computer designed for automatic translation, the exterior isomorphism obviously has no importance and nobody would expect the automatic translator to look like a brain or a human being.

Furthermore, what was previously said about anthropomorphic robots becomes useful: the similarity, at a certain observation level, between the exemplar and the artificial object or device, is often important only in interfacing terms, i.e., to make the machine *human-like* or at least *friendly* from the standpoint of the person using or, better stated, interacting with it. This fact tacitly assumes that such interfaces, which actually increase the chance of deceit, could make it easier to interact with the naturoid, overcoming people's difficulty in accepting the idea, often frightening, that the artificial has its own 'nature'.

Finally, we must keep in mind the fact that even among naturoids, processes or machines which are concrete and analytical, very often the analysis of the exemplar's structure does not lead to the decision to reproduce it as such—either because of its complexity, or because of its intrinsic non-reproducibility given existing conventional technologies—whereas it is considered possible to proceed with the direct reproduction of its essential performance. In these circumstances we can say that researchers and designers provide an applicable example of the principle of *functional equivalence* or *equifinality*. This principle—different from the similar concept in the Einsteinian theory of relativity, but in a weak sense—comes from biology (von Bertalanffy 1968) but also had applications in human sciences (Luhmann 2007), and states that it is not rare, in nature or in social phenomena, to find that a certain function can be carried out by a different structure than the one originally intended for such a task. This is the case, in the technological field, of a digital circuit based on electromechanical parts replaced by another based on integrated, solid-state logic circuits, or, in the sociological field, a social institution trying to provide welfare to orphan children by substituting the family.

Understandably, the principle of equivalence is also at the roots of the hopes of artificial intelligence and robotics designers since, having clearly to deal with materials and architectures very different from natural things, they claim that the processes of the human mind do not need a human brain to be realized. To this aim is devoted the well-known Turing test, for establishing whether or not a software works intelligently—that is to say, like a human mind—based on the equivalence and the indistinguishability of its result compared to the one generated by man. The equivalence principle can surely be applied effectively everywhere, and particularly within technological areas since many performances can be produced by means of very different strategies, starting from the many ways of generating energy. Nevertheless, when the core of the project concerns a matter which is intimately unknown, the fact that we get an outcome which is similar at one observation level, does not imply that the core of the process has been captured. In the field of artificial intelligence this has been discussed many times in relation to the Turing test. In this regard J. Searle (1980) proposed his famous argument of the Chinese room: if a man who does not understand Chinese is equipped with a way to transform English words into Chinese ones, he would be able to translate any message, while continuing to ignore the Chinese. On the other hand,

> The 'relation to a content' as characteristic of any intentional act implies not only the *extensional* reference to names of objects, but also *intension* of a conscious significance by which we associate names and objects in different contexts (Basti 2001)

This means that the Turing test privileges the syntactical observation level, quite neglecting the semantic one, a selective decision that is consistent with our views on any design of naturoids and which also reflects the current state of the art in the artificial intelligence research area.

When naturoids work in a satisfying way, we can assert that a high degree of functional equivalence is always present—and for this reason it is not useful in distinguishing a particular class of artificial objects—for, by definition, the artificial requires the use of different materials than those used by nature. Nevertheless, although the models which guide the artificialist on the one hand and those that guide nature on the other, appear homologous at a certain observation level, it can always be shown that they will be different on all the other levels, with consequences that can be perceived in various aspects or qualities of the same essential performance. This fact, which may be interpreted as the usual and intrinsic transfiguration of any design of a naturoid, explains why, usually, the naturoid, though it took hint from a natural exemplar, undergoes subsequent developments that bring it far from the natural instance. The examples from computer science and artificial intelligence abound, of course, but it is a matter of a

Table 11.1 Classification of naturoids

	Concrete (material devices or objects)	Abstract (informational devices or processes)
Analytical (reproduction of structures)	Organs, cells and tissues, robots, Virtual Reality (if interfaced with the real world), miscellaneous (e.g., diamonds, grass, horizon, etc.)	AI (artificial intelligence programs), ANN (artificial neural networks), ALife (artificial life), GA (genetic algorithms)
Aesthetic (reproduction of appearance)	Sculpture, architecture, imitation gadgets, reconstructions, counterfeits	Drawing and graphics, maps, figurative arts, simulations, descriptive virtual reality or virtual environment

rather general trend. For example, while even the first radar devices in some way reproduced the ability of bats to orient themselves by means of ultrasound, OTH (over-the-horizon) radar is able to detect objects beyond the horizon, which no animal is capable of doing. Similarly, an artificial nose can detect scents and smell much more precisely than the corresponding human organ.

In conclusion, the intersection between the two dimensions which we have singled out—*abstract-concrete* and *analytical-aesthetic*—allows us to formulate a classification system which substantially covers all the possible cases of naturoids. The artificial objects or processes which are mainly characterized by two extremes of the dimensions under consideration will be placed in the classes which are created in such a way, as shown in Table 11.1. For example, the functionality which is necessary for an artificial organ suggests that it be placed in the class of concrete-analytical artificial objects or processes, whereas a doll or a puppet will be placed in the class of concrete-aesthetic artificial objects or processes, and so on.

The classification system we suggest does not only have the useful purpose of organizing a rather wide range of activities, objects, processes or machines otherwise assumed as undifferentiated. In reality, the main benefit of such a system—as in every other case of scientific taxonomy—is that it makes it possible to deduce of the general characteristics which a naturoid has according to its placement in one class or another. Therefore, we are dealing with a methodological instrument of remarkable potential importance, obviously provided that the individual classes are carefully recognized. In particular, it is plausible to claim that the constraints imposed by the selection of the observation level of the exemplar and of the essential performance, and the problems which concern the reproduction of several performances, the principle of inheritance and the transfiguration processes, as characteristics of any naturoid, assume however different and specific features in the four classes which we have identified. This is a reality which the theory we are proposing has only just begun to investigate.

Chapter 12
Automatisms and Naturoids

There is a region of the "naturoids kingdom", called the region of automatisms, which is found at the boundary between conventional technology and the technology of naturoids. We can find examples of automatisms dating back to ancient times, including the Egyptian technology of the pyramids. What dominates in this area is what we could define as the principle of substitution, the substitution of a technological device for actions once carried out by man.

A very clear illustrative case is the invention of the throttle valve for steam intake and exhaust in machines invented by Thomas Newcomen in 1812. The valve had to be opened and closed by a person, who had to pull the cord and release it at the right moment. According to various accounts, the person who was assigned to the valve, probably in order to avoid such a monotonous job, had the bright idea of tying the cord to the piston, so that the latter could pull the cord and release it.

As we know, the self-regulation which we are talking about was later improved by James Watt using much more complex and efficient devices. However, the main problem we are interested in is related to the actual artificial nature of these kinds of inventions. The problem concerns the conceptual difference between the reproduction of something and its substitution with something else on the basis of some new model. Automatisms almost always substitute human actions with a technological device which yields the same results (or better results). As such, they are not intended in any way to reproduce either the exemplar (normally man or one of his physiological sub-systems) or the natural performance which makes that action possible.

In short, we could say that the people who design automatisms, even though they normally concentrate their efforts on man's essential performance, are interested in its faithful reproduction only to the extent that the actions which that essential performance makes possible are effectively reproduced. For example, the device which automatically opens a department store door when clients enter and exit could be defined as an 'artificial porter' only by isolating the effect of the

M. Negrotti, *The Reality of the Artificial*, Studies in Applied Philosophy, Epistemology and Rational Ethics 4, DOI: 10.1007/978-3-642-29679-6_12,
© Springer-Verlag Berlin Heidelberg 2012

essential performance (opening and closing the door) but certainly not by the designer's effort to reproduce human vision of moving objects and the related reasoning.

Likewise, *brain wheels* (gears moved by *cams*) which made the mechanical programming of machine tools possible during the last century, had the purpose of automating machine tool's work by substituting man's action but without any reproductive aim of the human mental and physiological performances which make those actions possible. The same can be said for the thermostat which activates or shuts down a radiator or refrigerator, the gear that automatically shifts speed ratios in a car, the circuit that stabilizes a television set and, generally, all self-regulation devices based on positive or negative *feedbacks*.

A computer, programmed to administer a firm's accounts, can be included in this category since the modalities it uses to perform calculations have nothing to do with those adopted by man, just as the physical structure on which it is based has nothing in common with the human brain. The overall architecture of such kinds of software reproduce the human way of realizing accounts—and, therefore, the result is equivalent to man's reasoning—but the calculation is the actual objective to be achieved, and, to get it, all possible software and hardware procedures are instantiated regardless of their similarity with what happens in a man's mind.

In the field of artificial intelligence, the attempt to program a computer so that it produces stories or translates from one language to another has resorted for a long time to the effort to design a human model which could take into close consideration the essential performances that are involved in the human mind. However, even translating machines based on models which are deliberately independent of the structure of our mind are not only conceivable, but actually more efficient than the ones inspired by the first models of artificial intelligence, which insisted on the discovery of the human way of performing a translation in order to reproduce it. Definitely, an automatism consists of a device that is intended to reproduce some human action, paying much more attention to exploiting the right technology and the most efficient resources, rather than to trying to reproduce what happens in the human organism. Simply, an automatism neglects the exemplar in order to concentrate on and reproduce the essential performance as such.

Both in the case of naturoids in the strict sense of the word and in the case of automatisms, the final resulting action which results from the device can be 'deceiving' to whoever interacts with it: a correct commercial invoice does not allow us to know if it was made out by a computer or a human being just as a heart does not know if the impulses for its rhythm come to it from the natural organism or from an electronic *pacemaker*. Nevertheless, the distinction between a naturoid and an automatic device is of great importance. A naturoid is always focused on natural performances which, often, can lead to different actions in different circumstances, according to reflex or decision processes which require some model of reacting or reasoning that, in turn, require taking into account the human mind as an exemplar. On the contrary, automatism usually substitutes the final action directly, and therefore, could be viewed as a sort of pragmatic artificial. In other words again, an automatism incorporates an action in itself, rather than the natural cause that produces it.

It is clear that the reproduction of a hand muscle and its natural physical essential performance, which allows a patient to work in the real world, within certain limits, as he pleases, is a much different thing than the substitution of human actions in a mechanical automatism which accepts paper money in a change machine. The same is true, in short, for the analytical reproduction of any natural exemplar in which the essential performance is considered more important than the specific action which it makes possible. In this sense, a robot for generic uses, not necessarily of anthropomorphic kind—although today we are still far from being able to give it a sufficiently broad generic intelligence—constitutes a clear example of a naturoid, whereas an elevator which is able to move skillfully up and down in a building, stopping at the right floor without any liftman, has to be considered an automatism.

One of the characteristics, though not the most important one, which differentiates a naturoid in the strict sense of the word from an automatism, i.e., reproduction from substitution, is, therefore, the degree of flexibility involved. The greater it is, the more legitimate it is to speak of naturoids, since, in order to give a device greater flexibility, we inevitably must go from final action models to more and more generalized models of natural high-level performances which, for example in human behavior, make that action feasible. This implies resorting to so-called 'reverse engineering' (Dennett 1994) which, in our case, means investigating what happens in our brain when we touch or see an object in order to transform a pure automatism, like a surface or light sensor, in a nature-like device.

It is under these conditions that, in artificial intelligence, we have experienced and are experiencing the problem of the shift from *micro worlds* to more general models. Micro worlds are local models of the world which, for example, describe the typical situation found in a restaurant with waiters, customers, chefs, tables, the coming and going of people, and so on. By communicating this micro world to a computer, by means of appropriate programming, it has been shown that you can get adequately intelligent answers from the machine to questions that regard rather broad classes of events that can happen in a restaurant. The same can be accomplished, naturally, in other micro worlds, for example, supermarkets. However, it has also been proven, as we have already noted regarding the problem of the synthesis of observation levels, that the intelligence tested in one micro world is not interchangeable with the intelligence produced in another one, except if we build a new, and not so 'micro', model of the world. Without reproducing generic intelligence as a natural performance—and therefore also so-called 'common sense'—we would thus be able to reach dimensions of the problem that are definitely beyond our capabilities.

However, the most important characteristic which distinguishes naturoids from automatisms is the former's close necessary link with nature. In other words, a naturoid in the strict sense of the word, once again confirms its heavy dependence on nature—even though it subsequently tends to transfigure nature—whereas automatisms possess a much more marked technological conventional tendency from the very beginning.

Although we are dealing with a distinction which can turn out to be very subtle—since any naturoid, as we have emphasized many times, requires conventional technology—it allows us once again to recognize the two basic aspects of technology, from a motivational and an intentional standpoint. The pragmatic aim of technology, the control of nature, is in fact pursued by means of pure inventions which try to produce the desired effects, without any reproductive concerns. The reproductive aim of the technology of naturoids, on the other hand, can only be pursued using strategies which give primary importance to the exemplar and its performances as such. In short, naturoids come always from the effort to get a high degree of homology (similar structures) or analogy (similar functions or relationships) with the exemplar and its performance, while automatism does not take this into consideration. For example, the compatibility of the materials and mechanisms of an artificial limb with the rest of the organism—its homology or at least its analogy with the natural model and its essential performance—becomes an integral part of such a project; on the other hand, if the pragmatic aim is only to allow the patient to move a bit, any other external mechanical device, almost completely neglecting the anatomy and physiology of the limb in question—such as a wheelchair—will suffice.

In this sense, the technology of naturoids can be considered a field that is very closely related to basic scientific research, because, in order to advance, it needs knowledge and models of the exemplars and their performances while, at the same time, it can contribute, at least hypothetically, to the advancement of this knowledge. In contrast, the technology of automatisms, while sharing naturoid technology's goal of substituting nature or, more often, human actions, cannot contribute in any way to the scientific knowledge of the exemplars or their performances, precisely because, in order to advance, it does not consider their intrinsic features and behavior as they appear in the natural world.

Finally, whereas the technology of naturoids in the strict sense of the word shows the human desire to recreate nature—an undertaking which obviously includes the pretense of having an excellent knowledge of nature—conventional technology and automation technology are evidence of man's desire and capacity to control things and events in the natural world, by getting or without getting ideas from their way of being in nature.

To sum up, every technological device which can actually be defined as a naturoid acts as a substitute for something natural, while it is not true that a device which aims to substitute something natural is, in itself, something which tries to reproduce natural things in the same way a naturoid does. We must keep in mind, in fact, that many products of the technology of naturoids have nothing in common with the world of automatisms, although they involve the substitution of a natural object or process with something technological. For example, the domes in contemporary Japanese architecture (huge constructions that reproduce, among other things, alpine or tropical landscapes in which people can go on vacations) are certainly naturoids and, as such, substitute their own natural exemplars. However, they do not constitute the automation of anything. The same can be said of various artificial organs which are implanted into the human organism, and substitute one of its performances, as well as many drugs which are often actually developed as imitations, or even replications, of natural substances.

Chapter 13
Naturoids in Real Contexts: Bionic Man and Robots

An artificialist's greatest aspiration, needless to say, is the reproduction of man. Marvin Minsky, among others, maintains that exploiting robot and artificial intelligence technology in order to repair our body,

> will be making ourselves into machines (Minsky 1994)

reducing our identity to a simple, slow and pleasant evolution. Others, like Hans Moravec (1989), Ray Kurzweil (2000) and Bill Joy (2000), predict that robots and AI programs will even supersede humans, menacing our species or, at least, making our nature irrelevant. Such kinds of prophecies seem to place themselves in the ancient tradition that saw in technological inventions or advancements nothing more than 'devilries', and represent very well the obstinate inclination to think that machines will autonomously reproduce themselves and will exhibit their own objectives, as humans do. In light of what has been said in the previous sections, it is very difficult to agree with such positions, many of which have already been exposed, by the way, in the central chapters of *Erewhon*, by S. Butler. Written in 1872, the book deals with the consciousness, and supremacy of machines, and related fears.

For instance, Butler asked

> Are we not ourselves creating our successors in the supremacy of the earth? Daily adding to the beauty and delicacy of their organisation, daily giving them greater skill and supplying more and more of that self-regulating, self-acting power which will be better than any intellect? (Butler 1872)

The prophecies that we have sketched above, which can be amusing on a literary or fictional level, seem to be like the fear of barbarians, or for any novelty that seems to menace our species and our life as it is 'here and now', neglecting the fundamental rule of coevolution which characterizes every field interactions, be it purely biological, cultural or technological. As a matter of fact, our way of looking at the world, and at ourselves, has changed many times through history, thanks to

M. Negrotti, *The Reality of the Artificial*, Studies in Applied Philosophy, Epistemology and Rational Ethics 4, DOI: 10.1007/978-3-642-29679-6_13,
© Springer-Verlag Berlin Heidelberg 2012

religious, philosophical and political movements. In the last two centuries, our cultural premises have changed much more in relation to the new advancements of science, than to the technological novelties of naturoids.

On the other hand, though sharing the same predictions, transhumanists think that the time has come to augment our mind and our body by means of technological devices. In doing so, they quite disregard both the unfeasibility of several purely imagined machines, and the fact that many of today's technological projects, though much more refined, have been the dream, or the nightmare, of the whole history of technology. Anyway, as has been wisely written,

> The field of artificial intelligence has repeatedly disappointed expectations ... History is littered with failed technologies once billed as inevitable. Some new technologies make it. Most do not. (Seidensticker 2006)

However, as we have seen in previous sections, such predictions, besides being based on advancements that today are quite unfeasible, also have several methodologically vague aspects which end up becoming real obstacles, in principle, in the realization of the task. The main point seems to regard the observation level: what does the word 'man' mean without any adjectives—such as anatomical, physiological, psychic, mental and social—that could define the dimension adopted for reproduction?

We already know that this is a problem we encounter in any artificialistic activity for, when an observation level is not explicitly established, the model remains vague, without any specification of the 'profile' we intend to select in order to observe and, subsequently, reproduce it. However, we also know that the choice of an observation level is an inevitable fact, of our natural habits, closely connected to our limits in interacting with the world. Almost all of the time, avoiding explicitly selecting an observation level means only selecting it implicitly.

This means that, even without any explicit indication by the designer, the observation level that he chooses, even unconsciously, will clearly appear in his concrete activity, i.e., in the aspects of the exemplar that he actually tries to reproduce. The adjective that is missing in the expression "reproduction of man" (or of any other natural exemplar) is filled in by the reality of the designer's actions.

Indeed, in the history of civilization, the real or imaginary cases of the reproduction of man are practically all oriented towards the implicit assumption of observation levels which we could call 'everyday', namely observation levels we adopt in our daily life through our common perception of things. Actually, this is the most ambitious choice, because an 'everyday' or generic man—unlike a man who plays a specific role, such as a physician or diver, a policeman or a customer—is a man who can think and act in many different, constantly changing ways according to heterogeneous situations. He has consciousness and common sense abilities as well as many other aspects. This is why contemporary anthropomorphic robots are usually designed on the basis of a much more narrow definition of man, almost always determined by some specific function or performance they have to accomplish. It is not by chance that, also today, the 'species' of robots abound in exactly the same figures that characterized the history of automata. We have now robots that play a violin or a piano, or even conduct an orchestra; robots who play with a man a given game; robots that walk, dance or press the keys of a keyboard and so on. These kinds of naturoids have a body made by electromechanical devices governed by a computer which processes the needed information, while the automata of the past centuries was made mainly of mechanical parts. Nevertheless, the final aim of both is the same: they have to be successful in performing a task of a given more or less wide class and only that one. In the end, they may astonish only people who had not participated in the design process.

Furthermore, there is another issue that joins current robots and classical automata, namely the solution that was found in the past centuries, and that is found today to give to the robot an exterior or human-like appearance. Clearly this involves problems that are not merely aesthetic, because, in the human body, the exterior appearance depends on the overall functionality of the organism. Therefore, the evident difference between the devices of a human body and those of a robot makes it very far-fetched to attempt to make up an external appearance of a robot as if it were a human being. The solution found in the past was the same used today for dolls: the automata was simply dressed like a man or a woman. Today's solution is, apart from a general aesthetically stylized design, sometimes the use of latex, whose properties are very similar, from a mechanical observation level, to those of human skin. In both cases, as we may expect, no real connection between the two observation levels, that is to say, the external surface and the internal system may be found because this is not the main performance to be achieved. The technological deception, this way, continues as always.

In ancient China, as Joseph Needham recounts, it is said that an inventor, Yen Shih, offered the king, who was visiting his town, an 'automaton' with features so real that the king himself had a hard time understanding what kind of gift it was since, next to Shih who was speaking to him about a gift, he only saw a man who he thought was the technician's assistant. Indeed, as Needham reports,

> …anybody would have mistaken him for a human being … (Needham 1975)

After learning that the man was, in reality, the gift in question and that he was an artificial realization, the king was astounded. The automaton, indeed was able to move, sing, look, and so on like any human being. Yen Shih even ran the risk of being condemned to death when the automaton began to court one of the concubines who accompanied the king. At this point, Shih showed him the 'pieces' that made up the automaton and the king was able to verify that it was a creature made of leather, wood, glue and lacquer, all painted in white, black, red and blue. Examining it more closely inside, the king recognized the liver, heart, lungs, spleen, kidneys, stomach and intestine and, over these organs, muscles and bones as well as the various limbs with their joints, skin, teeth and hair, all artificial of course.

By putting all the pieces together again, the automaton regained its initial appearance and behavior. By removing a certain organ, he lost the function which depended on it. The king was enthusiastic and thought that this might be proof that human ability could compete with the ability of the great Author of nature.

This fantasy story is at the core of current anthropomorphic roboticians, but it has no more likelihood today than thousands of years ago. This is due to all the methodological issues we have seen before, and, mainly, to the very hard problem to just 'putting together' two or more parts, both in terms of algorithms and in terms of physical parts. In fact, building a model that will preserve the properties and the requirements of the parts as standalone sub-systems would mean being able to govern those parts as it happens in the real organic exemplar. In turn, this would mean understanding what fine structures and processes allow connection and mutual adaptation between, say, the kidney and the liver or between the heart and the brain. Once we got such knowledge, we would be able to reproduce the communication and mutual adaptation in some new way that, very probably, cannot be reduced to a mechanical or informational process, as the winning technologies of a historical period may suggest. Anyway, whatever 'stuff' would characterize our model of connection and mutual adaptation between two or more sub-systems, it would inexorably affect the models of the two or more naturoids we had identified before as working well as standalone devices.

In other words, the coordination and mutual adaptation would indeed become the true new essential performance to be realized, putting in the background the exemplars and the essential performances that were selected for the former two stand-alone naturoids, and this could take the resulting whole naturoid far from the natural exemplar. This recursive outcome seems to be unavoidable in every bottom-up strategy because, working at a given level, we are forced to take into account, first of all, the requirements of that level, at the expense of the higher and the lower ones.

In Greece, as we recalled in the first chapter, mythology speaks of numerous 'living' creatures. We can also mention the Delphic oracles, which spoke through the wind or Talos, an automaton which Efesto built to watch over Crete. There is also the Hebrew legend of the Golem, a man of clay who was given life through letters of the "alphabet of 221 doors", on all of his organs. In this case, we have a very interesting *sui generis* reproduction, since the words are given the power of life and death over the Golem. Indeed, on his forehead the word *emet* (which means truth) had to be affixed, whereas in order to destroy him it was only necessary to eliminate the first letter, to get the word *met* (which means 'death').

Unlike the artificial imagined in the Chinese anecdote, the Golem is real in appearance but significant above all for his 'shift of state', from life to death. We find ourselves before a sort of 'programming' of the automaton, obtained through appropriate language and the right 'algorithm'. In the Renaissance, the doctrinaire tradition which dates back to the mythical Ermete Trismegisto, was reviewed by Marsilio Ficino and Pico della Mirandola. Together with alchemy, with its ability to transform one substance into another, this tradition favored attempts that aimed to reproduce life using mechanics. As Bruce Mazlish writes,

In the hermetic tradition of the Renaissance, the ancient charm inspired by automatons acquired new impulse. Magic and mechanics united intimately, and an air of amazement and fear rested on statues and angels which appeared on earth and in the air: are they real and living beings or not? Are mechanics, in turn, human beings, given that they are able to give life to whatever they build by imitating, in such a way, their own Creator? (Mazhlish 1995)

Through the centuries that followed, which saw Descartes' rationalistic lesson and Bacon's empiricism, up until today, the production of automatons or machines capable of reproducing a certain aspect of living beings, above all movement, has become a less mysterious endeavor mostly oriented towards science.

In the eighteenth century, Jacques Vaucanson appears to have been one of the more conscious designers of this new tradition. Even though, as M.G. Losano recalls, he was suspected of using tricks, his main purpose was not only to astonish, but, at least to a certain extent, to faithfully reproduce what happens in nature. For example, regarding the digestive system of his artificial duck, Vaucanson claimed that

… the food is digested in the same way as in real animals, by decomposition and not by trituration, as some physicists maintain (Losano 1990)

Note the following design by Vaucanson—described by Ceserani—presented by the French technician at the Academy of Lyon in 1741 for the

construction of an automaton which will imitate in its movements all life functions, blood circulation, respiration, digestion, muscle, tendon and nerve movements and so on. The author believes, by means of this automaton, that he will be able to experiment on animal functions and make inferences in order to acquire knowledge on the different phases of human health, in order to find a remedy for its illnesses. This ingenious machine, which will represent a human body, will be used in the end for a demonstration in an anatomy course (Ceserani 1969)

Thus, even here we can find the ancient ambition to rebuild a natural exemplar, 'putting together' many essential performances that appear at heterogeneous observation levels, but as usual, under the power and the limits of the sole, dominant mechanic technology.

The definition of ALife's goals, described by Langton in 1992, sounds significantly similar though it adopts a different technology, namely information processing that elaborates some kind of algorithms. It is particularly clear concerning the possibility, indicated by Langton, of

… reproducing the dynamics of life in other physical supports and thus making them accessible to new types of experimental manipulation and control (Langton 1992)

Though both Vaucanson and Langton are aware that they are operating on materials which are very different from those provided by nature for its living products, they still believe that it is possible to acquire knowledge on real life. Nevertheless, due to its formal stuff,

> Computational modeling (virtual life) can capture the formal principles of life, perhaps predict and explain it completely, but it can no more be alive than a virtual forest fire can be hot (Harnad 1994)

Likewise, the choice or the construction of a single observation level—anatomical, physiological or informational—regardless of the precision of the reproduction of structures and processes which are made accessible on that level, does not prevent the designers of naturoids from considering their designs as reproductions of natural exemplars, or from carrying out studies on these designs as if they were natural reality itself. The complexity of the natural exemplar, which cannot be reduced to one dimension or level, does not prevent both Vaucanson and, in more sophisticated terms but inevitably equally limited ones, ALife researchers, from following a more ingenuous principle according to which the *visibility of movement* (of a duck's muscles or of a 'colony of cells' on a computer monitor) is proof that reproduction has taken place.

However, it is interesting to note that the mechanical observation level, typical of the eighteenth century and consistent with the development of anatomy and physiology in that era, and the informational observation level, typical of today's development of informational theories and technologies, are quickly being replaced by more pragmatic observation levels when it comes to building concrete naturoids, such as industrial or military robots, or robots of any other kind.

In these cases, the "reproduction of man" very soon becomes the reproduction of some of his well-defined essential performances reproduced consciously without any noteworthy reference to what occurs in human beings, above all when a superior function is involved, such as recognition, control ability, typical problem solving, or calculation. Even in the field of robotics, therefore, it becomes clear that any naturoid involves a transfiguration of the exemplar and its performances. Moreover, robotics sometimes specifically aims to transfigure in order to pursue its own pragmatic goals. For example, a robot can be given the (certainly non-human) ability to make calculations or comparisons and rational decisions in environments full of gas or right in the middle of a battle; the power to rotate its wrist 360 degrees or to focus its hearing on a few frequencies and exclude others; and the ability to find in its memory, with 'inhuman' speed, geometrical shapes or very complex data configurations. Of course, the technology of naturoids can be very useful for scientific or educational purposes as well, as Vaucanson had sensed. Today, we use extremely useful artificial 'bodies' to test surgical techniques or seat belts, ergonomic models and various other types of devices.

However, the main ambitions of robotics remains an unreached—and intrinsically unreachable—dream if by anthropomorphic robot we mean even only a physical reproduction of the human organism, which can be operated with or on, as if it were a living 'general purpose' organism. With a few precautions, this reproduction may be possible, as we have said above, for specific purposes, i.e., where a performance of the human body, considered essential in some specific situation, is persuasively and efficiently isolatable, at some observation level, from the organic context of its whole, and sufficiently independent of its materials.

For example, so-called case-based robotics certainly seems to be a very promising approach which aims to give robots a more flexible behavioral intelligence more similar to human intelligence in certain categories of real situations. In these designs, the robot must be able to select the best solution from among those available, on the basis of assumptions and evaluations of the situation in which it must act.

Such a purpose will nevertheless be pursued with a great deal of pragmatism, namely compelling the machine to carry out the desired behaviors by means of software and hardware devices which constitute the 'tricks' and 'illusions' that are behind any naturoid. Thus, the case-based strategy can, for example, be combined with the object-oriented strategy, and therefore, as R. Bischoff, V. Graefe, and K.P. Wershofen write,

> … the robot will only recognize the objects which are important in order to carry out his operations correctly. Consistent with the object-oriented vision concept, the recognition is greatly reduced to a knowledge-based verification of objects or characteristics which are expected to be seen in the various situations (Bischoff et al. 1996)

Even robots controlled by neural systems (ANNs: artificial neural networks) are making their theoretical or prototypical appearance, making the best use of the speed and flexibility of neural systems while adapting to the configurations of the environment and its alterations. With these strategies we may be able to design 'artificial creatures': systems that are able to perceive the events of the outside world with great precision and act accordingly by means of real-time planning of the actions to be carried out; many think that these naturoids will also be able to learn from experience, and possess some kind of motivation, purpose, etc. The main problem lies in coordinating all these and other performances. The solution must include some human, prejudicial decisions on the observation level most appropriate for a given situation, and certainly preclude the tendency to set up overall models whose rules are not known for even the simplest animal's brain.

Chapter 14
The Challenge of Complexity

One of the most important conclusions that we reach using the theory of naturoids is that, given any exemplar, its faithful and overall reproduction is hindered, first of all, by the impossibility of describing it fully and faithfully.

This insurmountable obstacle, as we have seen, is a result of the selective character of our observation: we cannot consider and describe any kind of object by simultaneously taking into account all the observation levels available and, even less likely, all the possible levels, of course. However, this is precisely what we must do, though ideally, for a complete reproduction of the exemplar. At a microscopic observation level, for instance, things are arranged in such a fine way that, today, it is not possible to emulate them, because

> Chemistry and materials manufacturing technologies will be required to make structures of increasing complexity, hierarchically organized, and with a precision greater than what is possible today. Biological synthesis of such materials may hold the key to new routes that might be utilized for future technologies. For example, the silica shell of a diatom illustrates the complexity of material synthesis that biological synthesis can achieve. Syntheses such as the silica shell of a diatom are remarkable. First, the precision with which the nanoscale structure of these biological materials is formed in many cases exceeds that of present human engineering. Second, the conditions under which these synthesis occur are very mild. These are physiological, low-temperature, ambient pressure processes occurring at neutral pH without the use of caustic chemicals. This is in marked contrast to human manufacturing of silica materials (National Research Council, 2001)

Apart from the technological subtleness needed, the problem of observation levels may be widened if we consider, as exemplars, objects or processes from the natural world which we know to be not only very complex, but also very closely linked to their 'background'—man, but also any kind of animal, a flower, a pond, a plant, intelligence, life, evolution, etc. Having to describe one of these realities, we immediately realize the extreme arbitrariness of the operation, which will always depend on the selected observation level. We cannot seriously consider a strategy which aims to grasp such realities completely and simultaneously on all possible levels.

M. Negrotti, *The Reality of the Artificial*, Studies in Applied Philosophy, Epistemology and Rational Ethics 4, DOI: 10.1007/978-3-642-29679-6_14,
© Springer-Verlag Berlin Heidelberg 2012

On the other hand, when we examine the possibility of reproducing exemplars which seem to be 'parts' or sub-systems of whole objects (like the ones mentioned in the previous section), it seems to be a less difficult problem. Indeed, we cannot deny that the prospect of reproducing man's limbs or skin—although it is certainly not an amateurish task—is not as difficult as the reproduction of man in his entirety.

This is easily explained, since many sub-systems have already been defined and described for centuries, not only as objects of daily experience, but also anatomically and physiologically, i.e., by acquired observation levels which have become an integral part not only of science but also of common sense and general perception. Likewise, the essential performances of an organism's sub-systems appear to be clear and at times even 'obvious', above all, at a macroscopic physiological observation level: thus the muscles are levers, the heart is a pump, the liver is a filter, and so on. On one hand, this explains why artificial intelligence researchers and designers, trying to define the essential performances of intelligence, grapple with a thousand controversies, difficulties and mysteries due to the very general purpose nature of these kinds of exemplars. In reality, in our minds, there are no reinforced observation levels or there is no confirmed knowledge regarding its various functions. Thus, designers find themselves in the same situation that bioengineers would have found themselves in before acquiring knowledge, for example, of the physiology of the bloodstream or the function of cardiac valves, discovered in the seventeenth century by William Harvey, though, as far as intelligence is concerned, we cannot resort, up to now, to any direct empirical description of the brain structures which enable it.

On the other hand, the above considerations make it easy to understand why, in the field of concrete naturoids that reproduce sub-systems, designers advance with considerable speed and success. Nevertheless, we must bear in mind that, both those who work in uncertain, complex and abstract artificialistic fields—such as artificial intelligence or ALife researchers—and those who work in more established concrete fields—such as bioengineering—have never been dissuaded or discouraged in their work by the above-mentioned difficulties nor have any real reason to be. Indeed, in order to design some naturoid, the designer must only possess, first of all, a representation or, better, a shared model of the exemplar and its essential performance and, secondly, the materials and technical knowledge necessary for its reproduction.

There is obviously a crucial difference between a reproduction based on representations which, in turn, refer to scientifically confirmed information and models—and also, implicitly, to some preferential observation level—and a reproduction based on hypothetical or very subjective representations—perhaps without any sure indications with regard to observation. This difference lies in the fact that, in the latter case, since there is no common observation level, the naturoid will not be considered persuasive in spite of its possible effectiveness at a certain level. For example, the reproduction of human intelligence according to a model which describes intelligence as the formal ability to follow rules could lead to the construction of a very useful, but unconvincing device, for it is clear that a reliable model of human intelligence should have much more than the ability to follow pure formal rules at its core regardless of the fact that the model may turn out to be appropriate in special circumstances. In contrast, a widely shared observation level, together with the existence of confirmed knowledge, makes the field in which the designer works more certain, and the acceptance or rejection of the artificial object which he produces easier.

This long preamble has the sole purpose of confirming the very delicate role of our choices, before passing on to what we could call the most emblematic and dynamic field of research concerning the artificial, namely the field of artificial organs.

In this field, the work of bioengineers is becoming more and more fascinating and efficient but is often accompanied by problems which seem to increase precisely as knowledge advances. Some of the greatest difficulties emerge exactly when researchers are forced, by the new problems faced during experimentation, to consider observation levels, and thus essential performances of the exemplar, which were unexpected at the beginning.

This problem, as regards several types of biomaterials needed for artificial organs, was been summarized in 1995 in the following formula by Tirrell and Hoffman, two researchers from the Minnesota Biomedical Engineering Center

> … if we wish to predispose a material which has the characteristics of soft compound biomaterials, we must study the interactions which are involved at all levels: between molecules, up to the cells, up to the macroscopic characteristics of the tissues involved (Tirrell 1995, in Hoffman 1995)

Nevertheless, according to some researchers, improving our knowledge of what is going on at the microscopic observation level could not be enough to reproduce it, for

> … implantable materials with very fine mechanical and structural properties for host-cell migration and proliferation in order to create new hybrid artificial organs or tissue-engineered systems cannot be produced from synthetic materials. Biological materials have an extremely fine structure and unique properties that cannot be imitated with synthetic polymer materials (Noishiki and Miyata 2008)

Chapter 15
Illusion and Compatibility

Every organized system defends its boundaries in order to safeguard its identity, and biological systems are masters at this. In this sense we can assert that the theme of biocompatibility and biofunctionality sums up well a part of our comments regarding the difficult and perhaps, beyond certain limits, prohibitive ambition of reproducing natural objects—for example, human beings—capable of engaging in normal relationships with the rest of the environment or organism.

Biocompatibility can be defined as a property of surfaces. It is here, in fact, that an organism recognizes, and thus accepts, or does not accept, any substance it comes in contact with. In other words, the interaction between an artificial organ and the biological environment of the organism begins precisely at the surfaces of tissues and cells.

According to several researchers from The University of North Carolina who are involved in the reproduction of the esophagus, as in all cases where a natural organ is replaced by a naturoid the end result is that complications which can be fatal often develop. Complications associated with the implantation of an artificial esophagus include: anastomotic infiltration, infection, erosion, stenosis and displacement of the prosthesis. Infiltration and infection develop when there are biocompatibility problems between the prosthesis and the host organism.

Since even the use of bioinactive materials—which should not interact significantly with the organism—has proven to cause complications, researchers are now working on bioactive materials which, however, do not offer the same advantages as inactive materials. Hence, a hybrid artificial esophagus has been designed. It is partly based on traditional, conventional materials and technologies and partly based on biomaterials, or special conventional materials which display sufficient compatibility with the organism. The device consists of an internal silicon tube and an external tube made of a sponge of hygienic collagen. The collagen has been applied to promote regeneration of the tissue, especially the epithelium.

M. Negrotti, *The Reality of the Artificial*, Studies in Applied Philosophy, Epistemology and Rational Ethics 4, DOI: 10.1007/978-3-642-29679-6_15,

© Springer-Verlag Berlin Heidelberg 2012

The silicone tube is thus soon covered with natural tissue and, once regeneration has taken place, it can be removed because the cells do not adhere to the artificial esophagus, since silicone is one of the most bioinactive materials. The exemplar and its essential performance are therefore only reproduced artificially *pro tempore*, protecting the organism's boundaries and leading it, at the same time, to self-reproduce the organ towards which there will no longer be any immune defense response (Takimoto, 2000).

A similar strategy is also followed for artificial bones. As an expert, Jillian E. Cooke, wrote,

> A bone is a fascinating 'nanocompound' [composed of extremely small structures; Editorial Note] object: it is hard and strong at the same time. Its properties have proven to be extremely difficult to reproduce using conventional materials. Recently, researchers have synthesized a ceramic-organic nanocompound using strategies which imitate the way in which the organism synthesizes a natural bone (Cooke 2000)

Within the framework of a hybrid strategy, in fact, a 'matrix' is adopted—a biomaterial structure allowing cell growth—made of a compound called apatite organ, which acts as a 'root' for cell action and bone growth. *In vivo* experiments conducted on rabbits at the University of Illinois have demonstrated the reabsorption of the artificial interface, followed by regenerative processes of the femur.

This is a technology which still has many uncertain aspects, but which seems promising. According to the Laboratory for Surface Science and Technology (LSST) in Zurich, the two main aims in the study of the cases of biomaterials on surfaces consist, first of all, in a deeper knowledge of the superficial processes which come into play when a biomaterial comes into contact with an *in vitro* biological environment. Here the problems concern the alterations of the composition and structure of oxide films, the adsorption (the process of accumulation of molecules on an interface) of biomolecules, and the interaction with cells.

The second aim consists in modifying the biomaterial surface in order to obtain specific properties and improve the biocompatibility and biological functionality of the material that constitutes the naturoid. Regarding the materials used, research is now concentrating on titanium and its alloys in the market.

The physical boundaries between artificial and natural in the field of artificial organs are so important that the International Organization for Standardization (ISO) and the American Federal Drug Administration (FDA) establish very clear typologies in order to provide a useful classification for biological compatibility tests. For example, ISO establishes three basic categories: devices which are placed on the surface, external devices which communicate with the organism and implantations.

The case of artificial skin, introduced at first in the 1980s by Yannas and Burke (respectively of the Trauma Services at Massachusetts General Hospital, and chemistry professor at the Massachusetts Institute of Technology), is, obviously, especially crucial in this framework since it simultaneously concerns the first and third category of devices. Today, various products exist in this area. Two devices have been approved by the FDA: Integra—Artificial Skin Dermal Regeneration

Template—and Original BioBrane. Integra is a two-layer membrane, the first layer comes from bovine collagen and from a manufactured substance (*glycosaminoglycan*) that presents a regulated porosity and a well-defined degradation rate.

The temporary epidermal substitute is made of a synthetic (silicone) polymer and acts to control the loss of humor caused by a burn. The infiltration of fibroblasts, macrophages, lymphocytes and capillaries which come from the bed of the burn are generated through the first layer. As the healing progresses, the fibroblasts deposit an endogenous matrix of collagen and the layer of artificial skin is degraded.

In a subsequent stage, after an adequate vascularization of the dermal layer—and with tissue taken from the same patient (autograft)—the silicone layer is removed and the thin layer of epidermal autograft is placed on the 'neodermis'. The cells of the autograft will grow and form the corneal layer, closing the burn and making the functions of the dermis and epidermis once again possible.

However, the strategies described above are not typical of the world of bioengineering. Indeed, the concept of biocompatibility constitutes a particular case in a wider field which concerns the absolutely general problem of matching naturoids and nature. This matching is based on the need to control the transfigurations that the artificial could set off because of the principle of inheritance and the choices it comes from (observation level, exemplar and essential performance).

An insightful American writer, Ken Sanes, has dealt with the strategies adopted by the designers and creators of artificial environments or landscapes (*artificial naturescapes*) created to 'deceive' people, for example, visitors at an advanced technology zoo. Naturally their first task is to carefully observe, directly and indirectly, the environment in which the exemplars of their artificialization task are found.

For the Lied Jungle project, in Omaha, Nebraska, many specialists from the construction companies involved visited Costa Rica where they thoroughly studied the characteristics of the trees necessary to reproduce the forest. The main operation consisted in obtaining the external shape of the bark by means of latex layers. After the latex had dried up, they had the tree trunk's imprint.

Once transferred to Lied Jungle, the imprints were pressed on appropriately shaped concrete columns measuring 15–24 m, thus providing a formal 'replica' of the bark. The same technique was used to reproduce the rocks, which were touched up by hand, as well as the plants, using colors which were as close to the natural ones as possible.

When this stage was completed, a few natural plants were mixed with the artificial ones and the result is that the visitors, while admiring such an organized landscape, lose their ability to distinguish natural objects from naturoids just from the 'surface'.

In Jungle World, another landscape installation, the visitor, by standing in a certain position, can look at two layers of leaves: the leaves of the first layer are made of polyester, while the second ones are rubber plant leaves, and it is impossible to distinguish between them.

We now move on to another observation level, the acoustic level audible to man. In Jungle World, loudspeakers hidden in the trees continuously emit the chorus of insects and birds, recorded in a Thai forest, which is mixed with the sounds produced by the animals in cages.

The strategic requirement of invisibility, which protects the natural on the one hand (the audience which must be deceived) and naturoids on the other hand (the illusory devices), is thus met according to a logic which is perfectly analogous to the one that bioengineers follow in their task of separating organisms and artificial devices.

Indeed, up to this point, Sanes comments,

> … we have only begun to scratch the surface of illusion. If we deepen our investigation, we will discover that these 'immersion landscapes' are also based on secret simulations, that is, those that use a partial invisibility, cover-ups and distracting events in order to hide characteristics which could interfere with the illusion. As in every good trick of magic, art is what the audience cannot see and what it does not know (Sanes 1998)

Rightly, Sanes emphasizes that,

> … with a few variations, these characteristics can be found in all 'worlds' invented by man and which become part of popular culture: from theme parks to films to virtual reality machines (ibidem)

Furthermore, to avoid contact between organic elements (zoo animals) which would be, so to speak, 'rejected' by the host system (the audience), while allowing the audience to enjoy the view in a realistic environment (the true essential performance of the whole installation), a whole series of devices is used. It includes artificial rocks and islands on which the animals can be observed and, at the same time, isolated, thin electrical wires and even Vaseline on tree trunks and branches to prevent the monkeys from climbing up the trees.

Here we also have a considerable contribution which comes, as we might expect, from conventional or artificial technology. Air conditioning ducts, drainage and irrigation lines, land supports, feeders and everything necessary to facilitate the sensation of being in a tropical forest are well hidden in rocks and trees or behind other obstacles.

Understandably, not everything is hidden and many things are completely excluded, such as the sights, sounds and weather which should surround the environment. Other things, however—such as side effects which inevitably accompany naturoids or infections which affect artificial organs—occur unexpectedly and create difficulties for the installation. The intrusion of some forms of unauthorized wildlife is a particularly common problem. These intruders include mice and cockroaches, which enjoy the artificial forest habitats, with their ideal climate and abundance of food and places to hide. Hence, caretakers are constantly at work not only to keep a few natural organisms alive, i.e., the zoo animals, but also to eliminate other forms of life which get into the crevices of the constructions.

In short, the caretakers act as highly selective drugs or antibodies to 'regulate' the processes which are considered essential. Here the problem of compatibility takes on the form of man's careful and intentional selection of the animal species that will be authorized to survive in the zoo, in order to prevent the development of a completely natural food chain, one which is harmful to the installation's purposes. Thus, toucans have been removed from Lied Jungle because their behavior, which includes destroying habitats, taking away other birds' babies, and eating various kinds of frogs and lizards, is considered unacceptable.

As in many bioengineering cases and, in the end, in all naturoid objects or processes, 'deception' and illusion rely on expensive and complicated monitoring and control systems, which aim to hide and protect something while allowing something else to be activated for the achievement of the essential performance.

The problem is even more real in cases in which man tries to get nature to reproduce itself by setting up the right conditions, partially artificial and partially conventional, in which nature needs to function. While this is becoming a real possibility, starting from the deepest structure of biological systems, i.e., the DNA, it is much more difficult where the open, natural environment is concerned. These kinds of difficulties were encountered in Texas, where man has tried to recreate several wetlands for environmental purposes. According to the *Texas Water Resources*

> Some studies show that as many as half of all created wetlands fail to achieve desired goals. Concerns revolve around such issues as the complexity of reproducing natural systems, the difficulty of measuring the success of man-made wetlands, the ability to mimic wetland functions such as flood control or water quality improvement, the extent that aquatic life will utilize the sites, and long-term success (TWR 1992)

Indeed, in these situations, the task of establishing the boundaries of the exemplar is very difficult, and the relationships among the many observation levels which are involved become intricate and largely uncontrolled.

In conclusion, it should be noted that, in addition to the already mentioned Japanese domes, the history of architecture is full of examples of artificial land scapes. One of the most important examples can certainly be found in the twelfth century when the doge of Venice, Caprese, had the architect Nero Faggioli (founder of the School of Lattuga where great masters such as Brunelleschi and Ghiberti were educated) build him an artificial mountain landscape, complete with a garden, a zoo and even a stream which, driven by a pump, flowed down the mountain.

Contemporary landscape architecture, in turn, has developed very sophisticated models and techniques to match natural installations, like gardens, and the technological and organizational requirements of towns. Nevertheless, in imitating natural phenomena rather general and arbitrary features can be dangerous. Thus, it has been noted that

> Resolving an urban issue of accessibility with an abstract representation of nature generates all sorts of unexpected relationships. (Aben and de Wit 1999)

The search for the right model—and the related observation level—to adopt in order to capture the most plausible essential performance of a natural exemplar is always one of the main problems in the design of naturoids. To carry out good work, the relationships between the naturoid and the surrounding environment have to be carefully studied in advance in all cases, even in the most seemingly simple, like artificial grass, because

> Large areas such as the shock pads of artificial sports surfaces can experience very considerable dimensional changes due to temperature. In one particularly baffling case artificial grass appeared to expand and formed waves when the temperature was lowered. The effect was found to be a result of internal stresses and the differential expansion between the grass and the rubber shock pads (Brown 2002)

The key point is that, despite any care given to the rebuilding of the essential performance, in all cases, the range of the *natural effects* coming from a naturoid is potentially great, and they may affect even the scientific study of natural phenomena since

> … conclusions based on studies of birds at artificial sites might be artifactual. This is potentially a problem with any study of animals occupying human-altered habitat and is an especially serious issue with hole-nesting birds that now breed primarily in birdboxes often created by the scientists themselves (Brown and Brown 1996)

In the field of proto-cybernetics, at the beginning of the past century, many observers tried to study human high-level performances by building artificial devices based on this or that model, a theory or supposed analogy, like memory modeled as a relay or a hydraulic system studied as analogous to a nervous system. But, studying a naturoid on the basis of a model actually means studying the model rather than the natural exemplar. Thus, H.S. Jennings, discussing the theoretical machines designed by J. von Uexküll, wittily said

> From the properties of these machines it is not possible to conclude what physiological properties we will discover, because the parallelism is far from being complete, and, thus, there are two systems to be understood instead of one only (Cordeschi 1998)

Part V

Chapter 16
Naturoids: Interface and Camouflage

While naturoids always need to interface with nature through *ad hoc* devices, the naturoid itself can appear as an interface. This is the case for a whole series of objects or devices, which have been used in military camouflage. Such devices and objects are now coming out on the market, especially in the US. Their purpose is to protect local environments—for example a residence—from the surrounding environment. The same is true of the products of Larson Utility Camouflage in Tucson, Arizona. As its entrepreneurial mission, the company declares that it

> … wants to help keep utility devices, such as cellular telephone towers and transmission antennae, as inconspicuous as possible, minimizing visual intrusiveness on surrounding neighborhoods and natural environments. We conceal cellular transmission towers with artificial foliage and bark replicating any tree species, for example lodge pole pines, saguaro cacti and fan palm trees. Larson's architectural facades, crafted of fiberglass or lightweight concrete, can be designed to blend with virtually any building style, while shielding antennae, dishes, masts, poles and towers. Our artificial rock and plants, virtually indistinguishable from nature's own, serve as sound barriers and erosion control for highways (Larson Utility Camouflage 2001)

Such kinds of aesthetic naturoid can play a perceptive role in giving to some areas, affected by various types of man-made disorder, a more pleasant aspect. Nevertheless, since the above quotation comes from an advertisement, obviously there is no mention of the possible 'adverse side effects' of these products, i.e., the undesired effects or degradation which could occur once they are placed in permanent relationship with the natural environment.

In contrast, the techniques used in making 'artificial reefs' are based precisely on these effects. A man-made object is placed underwater—studies take advantage of the observation of sunken ships—and allowed to become a part of the ocean ecosystem. The host environment attacks it with its own forces and aptitudes in order to reduce it to something compatible with itself. For example, barracudas are known for their ability to extend their own territory to include a sunken ship just moments after it has sunk. One of the main goals of artificial reefs is, as Indiana

M. Negrotti, *The Reality of the Artificial*, Studies in Applied Philosophy, Epistemology and Rational Ethics 4, DOI: 10.1007/978-3-642-29679-6_16,
© Springer-Verlag Berlin Heidelberg 2012

University researcher's report, to draw increasing numbers of swimmers to these structures, in order to lighten man's pressure on the reefs in the natural ecosystem.

Furthermore, since natural reefs are getting rarer in many ocean areas, artificial reefs have also become important from an economic standpoint. They increase opportunities for recreational fishing and swimming sports in coastal waters, as well as the habitat's overall productivity. Steel or concrete structures are placed underwater, as well as pierced fiber-glass balloons, which will act as a support for the development of the communities of organisms that normally grow on natural reefs. In theory, and depending on the materials used, artificial reefs could be usable for several centuries. What is interesting here is that, according to Yehuda Benayahu of Tel Aviv University,

> … with time, there is an absolute similarity between natural and artificial habitats (Kern 2001)

This is quite consistent with our assumptions: indeed, the natural environment around the artificial habitat—which is intentionally designed without established boundaries—will interact with the device at all the possible observation levels and will eventually transform it according its own rules and requirements. Thus, nature completes, so to say, man's work, adding, with time, all the features which were not intentionally designed. In cases like this, the aptitude of nature to 'reduce to itself' any man-made object, is the true aim designers want to achieve.

On the other hand, this is a very special class of situations in which the emergence of properties does not come from the naturoid as a rationally designed product, but, rather, from its 'degradation' under attack from external natural phenomena which generate just what the designer desires. Something similar happened in a recent experiment on artificial leaves. The study aimed to understand if a particular kind of fungus, which lives on the surface of a particular kind of leaf, could colonize artificial leaves as well.

According to the researchers

> Colonization of artificial leaves also supports previous reports that epiphylls are not species specific in their use of substrate although host specificity may become important under conditions of stress, such as low water availability. Microhabitat humidity, light and nutrients transported by rainwater, dust, animals and falling micro debris apparently were enough for successful colonization of the artificial substrate (Monge-Nájera and Mario Blanco 2001)

Determining the differences between natural objects and naturoids becomes, in other cases, of utmost importance. For example, barriers needed

> … when reprocessing spent nuclear fuel are ultimately to be disposed of in deep underground rock formations. Designing safe and sensible disposal facilities and selecting their sites will require methods by which to accurately assess the performance of deep rock formations (natural barriers) and the glass logs (artificial barriers) encasing the HLW (High-Level Wastes) to be buried (Kawanishi 1995)

Of course, a different 'philosophy' comes into play when man tries to defend a landscape from some natural erosion phenomena by building *ad hoc* barriers. As has been remarked

> The public has long considered that artificial barrier dunes are the natural or desired configuration and that the erosion of the shoreline is detrimental. Application of the philosophy of adjusting to and living with the forces of nature will require new efforts to inform the public of the constructive nature of an every changing landscape where erosion of the coast plays a significance role in maintaining the environmental health of these areas (Stuska 1998)

Anyway, when anthropomorphic robots are adopted in interacting with people, the human-like aspect and behavior are viewed as a strategic interface and become a part of the essential performance the robot is built around. As a result, anyway, it has been found that even young people, such as students at a university campus, feel fundamental psychological differences communicating with this kind of naturoid as compared to humans. For instance, they give much more value to inconveniencing or hurting, and expressing disapproval, to humans than robots (Kim et al. 2009).

Chapter 17
Structure or Process?

The exemplar and essential performance always involve an unlimited number of observation levels, but, as we might imagine, designers can only consider the levels they have knowledge of or, from among these, the ones that seem more easily approachable from both a scientific and a reproductive standpoint.

As we have pointed out, every single design of a naturoid, whether it be analytical or aesthetic, concrete or abstract, *de facto* can only take into account one dominant observation level. There are circumstances in which this constraint seems to completely absorb researchers, as in the case of artificial intelligence. But also in the field of concrete naturoids the improvement of a design can lead to several subsequent shifts of observation levels, at times giving rise to designs that turn out to be more complex than they first seemed.

The case of artificial blood is typical, in this sense, because its extreme complexity is in no way inferior to that of other types of organs, tissues or biological processes. Artificial blood must have a suitable fluidity and the ability to provide nutritional supplements like natural blood and, above all—this is the true essential performance of current designs which involve a proper observation level—to transport oxygen. This is the essential performance of current projects. In fact,

> Blood does many things, of course, and artificial blood is designed to do only one of them: carry oxygen and carbon dioxide. No substitutes have yet been invented that can replace the other vital functions of blood: coagulation and immune defense (Winslow 2001)

In any case, in the field of naturoids, it is not unusual for the reproduced essential performance to show some improvement over the natural one. As a matter of fact, artificial blood cells

> … will perform their specialized function—delivery of oxygen to tissues—even better than blood (ibidem)

The loss of considerable amounts of blood can be compensated by the use of saline fluids which, however, do not possess the main property of blood, the ability

M. Negrotti, *The Reality of the Artificial*, Studies in Applied Philosophy, Epistemology and Rational Ethics 4, DOI: 10.1007/978-3-642-29679-6_17,

© Springer-Verlag Berlin Heidelberg 2012

to carry oxygen. At this point, the artificialistic strategies go in two different directions. On the one hand, there may be an attempt to produce compounds that do not carry oxygen directly, but which are nevertheless able to facilitate the solubility of oxygen in blood. In this case, there is a focus on the essential performance as such, regardless of the biochemical structure involved in the natural process, namely hemoglobin. On the other hand, research may aim to reproduce this structure directly, and therefore it becomes the real exemplar with its own observation level. Hence, researchers try to understand more thoroughly the finer aspects of the structure in question and how such a structure influences its predisposition towards oxygen.

Naturally, a common problem related to the two tendencies is toxicity. At any rate, the problems associated with the realization of artificial blood are closely related to the problems concerning artificial organs because, in many cases, blood is an inevitable interlocutor. In the case of the artificial heart, at least two physiological systems are particularly sensitive to implantation: the haemostatic (coagulation) and inflammatory systems. Their activation can lead to the formation of thrombi on the artificial device's surface and this, in turn, can cause its failure, an attack on the natural organ or a hemorrhage.

Hence, research is oriented towards the study of thrombotic and inflammatory processes in relation to the materials used in the implantation, while at the same time, it tries to provide treatments of the surfaces involved which minimize these undesired events and suggest improvements for the design of devices.

Even in bioengineering, designers' efforts may go suddenly beyond the exemplar and its features in the hope of finding *anyway* a good solution. Sometimes this means resorting to pure conventional technology and sometimes turning attention to other natural fields. This is the meaning of what happened recently to researchers at University College London, who

> … when confronting the problem of infections arising from the attachment of prosthetic limbs, claim to have found a suitable solution by turning their attention away from our species. They studied the processes underlying the periodic renewal of deer antlers, and, on the basis of the knowledge acquired, developed a titanium attachment device for the artificial limb that seems not to produce any infection (Sneddon 2008)

In the case of the artificial eye, on the other hand, the architecture of the natural retina, as a model, seems to be the only choice, since it is difficult to theorize a device able to perceive images from the outside world which does not have a dot matrix structure. As we have seen in a previous section, citing Mahowald and Mead's work, the reproduction of the retina is the main step towards building an artificial eye and it must comply with a few inevitable structural conditions while temporarily neglecting others.

The artificial eye as a structure must, in any case, be combined with the eye as a process, i.e., with what we call 'eyesight'. This process, an essential performance which prevails over all other concerns, is, however, reproduced under the constraint of the exemplar's structural nature, specifically the retina. In other words, since we must make sure that the retina sends electrical signals that correspond to what the brain—the true center of ‘seeing’—expects to receive, it must function under the control of an algorithm. The algorithm is carried out by a small computer or suitable piloting electronic circuit of the retina, which produces the desired signals in the most suitable form.

The essential performance of the retina, considered as the exemplar, must then make way for, or adapt itself to, the general essential performance’s requirements, i.e., sight as a process. This is, therefore, a good example of a situation in which naturoids very often find themselves: when the exemplar is viewed as made by several sub-systems, each of them could be conceived as a separate potential exemplar. In fact, there is always only one overall essential performance which must be fulfilled, and this ends up dominating the characteristics which the various sub-systems must assume.

An external artificial retina like the one designed by Mitsubishi Electric (Advanced Technology R&D Center) and Mitsubishi Electric Information Technology Center America (MELCO Research Laboratory) called *image sensor*, capable of recognizing a person’s head movements and transforming them into input for a computer, will be controlled by an algorithm (*real-time vision algorithm*) which is definitely different from the one that must guide the work of the retina in order to obtain useful signals for the human brain.

In any case, once again, at least ideally, a natural organ, the brain, will be ‘deceived’ by the artificial device: indeed, the brain will process the perceptive data “as if” they came from a natural retina, maybe spontaneously reorganizing its processing features to adapt itself to that of an artificial device. The same thing could be said of the computer connected to the Mitsubishi retina, which will accept the input ‘as if’ it came from someone typing in commands on a traditional keyboard or from the movements of a ‘mouse’.

The Japanese have recently presented the prototype of a machine, also based on the perceptive and processing ability of a variation of the artificial eye, which is able to translate the hand movements of an orchestra conductor into signals capable of modifying various acoustic parameters of an electronic synthesizer (volume, dynamics, color, etc.). This certainly calls to mind—but we will leave it to the reader to unravel the meaning of this analogy—a remark made by Johann Sebastian Bach and related by Köhler:

> … the organ? There is nothing special about it: all you have to do is press the right keys at the right moment and the instrument plays by itself … (Köhler 1776)

Canadian Biomech Designs Limited, in cooperation with the University of Alberta and Dendronic Decisions Limited, has recently come up with the design of its C-Leg, an artificial leg for people who have undergone a trauma or the amputation of this limb above the knee. The C-Leg, like many artificial limbs or organs which must fulfill aesthetic and functional requirements at the same time, is the partial analogy of an anthropomorphic robot, in the sense that it must appear aesthetically as a leg and, inside, be able to function as a natural leg. As always, the two observation level levels, aesthetic and functional, are in no way correlated with each other, nor could they be on an analytical level, i.e., the device cannot consider and reproduce all levels, from the epidermis to the osseous cells.

Indeed, the C-Leg privileges, as an essential performance, the leg's dynamic flexibility, i.e., its ability to support the person's body by adapting to the various real situations which he could find himself in. The exemplar, therefore, is the leg as a mechanical structure, but its essential performance, intelligent flexibility, dominates the design and the structure is reduced to a physically suitable support, which is light and resistant enough, and so on.

The main structure of the C-Leg has nothing to do with a natural leg, since it is equipped with a computer which gives the limb local intelligence whereas, in the natural case, even without excluding the possibility of local control circuits, the organ assigned to such a purpose is the central nervous system. The C-Leg computer controls a valve that stiffens or loosens the knee joint based on a continuous cascade of signals which tell it how much pressure is on the limb. Hence, the presence of obstacles, the loss of balance, and so on, can be deduced from these signals. In such circumstances, the artificial leg stiffens the knee and places itself in a 'stumble state' in an attempt to prevent a fall. The task of Dendronic Decisions is to design the most suitable software for control of the limb.

While it is easy to grasp the delicacy of such a design, and a whole series of possible side effects (not on a biological level, but on a mechanical one), we must also consider the fact that a person, equipped with this kind of artificial leg, could be able to adapt very well to its operation after a period of training, just as occurs in the use of many machines which, very often, are operated better by their users than their designers were able to imagine. As a psychological side effect, the person's ability to interpret and predict the artificial leg's behavior could in fact come into play, with him adapting to it to such an extent as to compensate for the exclusively local nature of its intelligence. This could start an interesting interaction in which two systems—the man and the artificial leg—attempt to adapt to each other.

It is interesting to note that this sort of interaction was already noted in 1991 by a group of researchers in the field of computer-based musical instruments. They considered a rather 'intriguing problem', namely the relationship between the ability of some experimental electronic instrument to adapt to the style of playing of human being at the same time that the human being tries to adapt himself to the instrument (Lee et al. 1991).

Unfortunately, the above kind of undesigned but promising effects have several symmetrical counterparts. In fact, naturoids that have to be adopted as external organs may trigger psychological effects that could evoke the general problem of the relationships between a naturoid and the context it has to work in, namely, in the case of organs, the human being. In the case of the artificial limb,

> Psychological issues can impede the acceptance of electrical power; however, some of the more common reasons precluding the use of electric power are the patient's refusal to wear such a device based on a prior experience or the lack thereof, misconceptions about the nature of such a control system, or simply a refusal to wear an interface which is fitted more intimately (Muzumdar 2004)

The case of the artificial hand is even more complex, if it is to have not only the aesthetic appearance of the various prostheses in existence, but real capabilities of articulation and movement (excluding the use of fingers for the moment). The Utah Arm and Hand System, for example, uses a control (*Sensitive Proportional Control*) which is in turn piloted by the myoelectrical signals of two muscles. This control makes the regulation of elbow, hand and wrist movements possible. It is a device that gives the person with the artificial hand the possibility to move the elbow and wrist at different speeds, unlike other prostheses which can be operated solely *on–off* logic, i.e., by activating a switch.

Small stainless electrodes are placed on the skin so as to register the voluntary activity of the two muscles, that is, their myoelectrical signals. These signals are then amplified and used to pilot the prostheses' movement. Everything, as always, is made aesthetically similar to a natural arm and hand.

What differentiates this type of artificial organ from the one seen above is the designers' decision to select, as an essential performance, not so much the elbow and wrist movements—already available in the external electrical drive *on–off* prostheses—but the control of movement itself. The hand, arm and wrist, as in the previous case, only have minimal structural similarity, apart from aesthetics, whereas the true exemplar is the biological circuit which, in nature, makes control of the limb as an essential performance possible.

The structure of this circuit, obviously simplified, is rather similar to the exemplar even if the 'degrees of freedom' given to the hand and elbow are subject to considerable restrictions because the signals, produced intentionally by the person, are nothing compared to the wealth of the natural organs' innervation and musculature which the amputee does not have.

It is clear that many problems are irrelevant when designing organs which will not be implanted in the human organism, but, rather, are destined for machines such as robots that do not have any direct physical connection with man. This is the case of artificial muscles designed by American SRI International, based on the use of films of electrorestrictive polymers (which contract or stretch under electrical tension), to be used by a small robot. Artificial muscles, according to SRI reports, are comparable to natural ones as far as performance is concerned, but are more efficient and, above all, faster.

However, research on artificial muscles also includes the possible transfiguration of the natural exemplar, in this case an amplyfying transfiguration, because in BAM project

> Shahinpoor's research documents state that the fibers are capable of holding four kilograms per square centimeter. A human biceps can lift a maximum of just over two kilograms per square centimeter (Gibbs 1998)

The natural exemplar's structure, the culmination of millions of years of evolution, is never overlooked by designers who may, however, give up its reproduction when it is too complex, unknown or made of materials which are impossible to reproduce with existing conventional technology. In these circumstances, researchers' efforts are directed to reproducing the exemplar's structure only in analogical terms. In a way, the strategy of nature is accepted, but its operating principles are modified to a varying extent in order to obtain the desired essential performance through the available technologies.

This is what is happening in the design of the artificial nose or, rather, of artificial smell. The natural nose possesses millions of receptors, made up of proteins placed on the surface of the cells which are able to capture the molecules of odors. Nevertheless, there are only 10,000 types of receptors, a number which is not sufficient to recognize all the odors that we are exposed to. The brain recognizes odors by examining the combinations of receptors that each particular odor activates. Since there are millions of possible combinations, the brain is capable of distinguishing the odor of virtually all existing molecules.

Researchers' attempts to reconstruct the sensible "repertory" of the natural nose had not gotten very far until very recently, when John Kauer and David Walt from Tufts University found a completely new approach, which rejected nature's model and, in this way, tried to capture the essential performance, as such, by reproducing the function of the exemplar. The approach consists in substituting the natural strategy based on chemical recognition with a strategy based on optical recognition. The artificial nose is made up of a system of 19 optical fibers covered with different fluorescent compounds containing a dye called *Nile Red*. When the molecules of a given smell hit the fiber cover, it emits a fluorescent light with a different intensity and wavelength from the one emitted by the nearby fiber and, therefore, every possible odor produces its own fluorescent spectrum.

At this point, the spectrum is transducted into electrical signals and these are coded so they can be interpreted by a digital neural network, i.e., a software system capable of recognizing the various configurations that the data can have, by classifying them into sufficiently homogenous 'types'.

This is an artificial device which, perhaps more clearly than others, re-proposes the criterion of functional equivalence which we mentioned in the chapter dedicated to the classification of naturoids. It would obviously be absurd to study the artificial nose in order to gain a better understanding of human smell, since the receptors' operating principle is completely different and the analysis of the information from the combinations of receptors is carried out on an informational basis (by the computer) and not on a biochemical basis as presumably occurs in the brain. Nevertheless, the debate remains open regarding the possibility that, at least, the 'logic' of the natural brain might be reproduced by explicit algorithms or by the self-organization of digital neural networks.

At any rate, the transfiguration of the exemplar's structure and of the ways in which it carries out its essential performance in nature allow the artificial nose to carry out tasks which are completely impossible for the human nose, such as monitoring chemical changes inside the human circulatory system.

The fact is that the need for more and more refined artificial sensors in the field of robotics makes the observation and knowledge (and therefore the modeling) of natural sensors ever more important. So, among other organizations, even a European research group coordinated by the DIP-INFM of the University of Genoa is developing a research project on natural and artificial sensors in order to enhance our knowledge of the first stages of information processing in visual, olfactory and auditory processes to improve the devices which robots must use.

The natural processes assumed as essential performances are phototransduction (for sight), chemiotransduction (for smell) and mechanical transduction (for hearing) in vertebrates, whereas the applicative objective is the design of autonomous 'navigation' strategies for mobile robots.

Significantly, the research also intends to

> … compare natural sensors and artificial ones in order to establish which characteristics of biological sensory processing can be useful for the designing of new artificial sensors or robots (DIP-INFM 1995).

an aim which is at the center of bionic programs. In the various cases which the project deals with, reference exemplars include, the fly and several amphibians

The artificial heart has also been the subject of several different reproduction strategies. The structure of the exemplar and its essential performance (the pumping of blood in the organism's circulatory system) carry out different roles in the various projects. At Kolff Laboratories in Salt Lake City, for example, and in many other research and designing places, the anatomical structure of the natural heart constitutes an important aspect, because the reproduction of the essential performance associated with it depends considerably on the reproduction, even if it is modified, of the various cardiac sub-systems: from atria to valves.

In contrast, privileging the essential performance provides a new example of innovation regarding to the exemplar's structure and also a specific aspect of the essential performance itself, namely the pulsating activity of the pumping of blood. This is the case for the *Streamliner* project of the McGowan Center at the University of Pittsburgh, which is expected to be realizable within just a few years. It is a device which, when completed,

> … will permanently substitute the majority of the natural heart's pumping functions (McGowan 1999)

The innovation consists in the fact that, instead of reproducing the usual essential performance by technical contracting devices capable of generating a rhythmic and therefore pulsating pumping, *Streamliner* is designed to pump continuously and produce a continuous flow thanks to an electrical engine which rotates at 10,000 rpm. Hence, the recipient of such an artificial heart will not have pulsations.

As researchers at the McGowan Center maintain,

> … although *Streamliner* is different from a natural heart, it will take advantage of the sturdiness—and absence of various weaknesses—of the biomaterials it will be made up of. For example, thanks to the way in which an electrical engine works, it is much easier to design a constant flow compact device rather than a *bellows type* which requires compressed air lines that can allow infective biological entities to enter the organism (ibidem)

There is considerable interest in this project, because the transfiguration of the exemplar, but above all its essential performance, will likely produce various side effects which are now unforeseeable, apart from a few psychological effects, which we can already imagine, given the absence of pulsations.

Regarding transfigurations or side effects deriving from the materials and procedures adopted in the design, it is important to observe how the artificial kidney—another great protagonist of bioengineering—is able to remove both bromide and iodine just like the natural kidney. But the artificial kidney can also distinguish between them, and thus remove them according to their gradient. Moreover, the artificial kidney is able to remove potassium in a very short period of time.

The complexity of the exemplar's structure and its performances is daunting at times, given our current knowledge and technological capabilities. Apart from the brain, the liver seems to be the most complex organ to approach and reproduce at the moment. It is a kind of chemical system that functions 24 h a day and is responsible for the production, storage, metabolism and distribution of a considerable amount of basic nutrients for the organism's health. Besides producing glucose and proteins and factors which help the blood to coagulate, the liver acts on waste products and transforms some of them into usable elements, while discharging the ones that are dangerous. At the same time, the liver produces vitamin A and stores vitamins A, D and B12. In short, a chemical firm would need a plant stretching many hectares to do a similar job.

Clearly, before such a wide spectrum of performances, the selection of one of them as essential is prohibitive yet the reproduction of all of them would require a model, to design and coordinate the interfaces, which is presently beyond our capabilities. This explains why it is better to resort to what we could call a mixed strategy for this organ: this is the case of the misnamed 'artificial liver' under study at Edinburgh Royal Infirmary and the University of Strathclyde, Glasgow, which uses pig cells. A similar project, in the development phase at Queen Elizabeth Hospital in Birmingham, uses residues of human liver cells derived from transplant operations.

Interest in the latter project stems from the fact that the plasma, sent to the device through a tube inserted in the groin, passes through a network of tiny permeable plastic tubes which force it to mix with the cells, thus removing toxins and converting the necessary compounds. A second network of tiny tubes then restores the treated plasma in the organism.

The composite nature of this object makes it a typical example of hybrid artificial, in which a direct attempt is made to match a technological sub-system, namely the network of tubes, with a completely natural one, namely the liver cells. The essential performance can proceed undisturbed by reactive phenomena in the organism because it is realized on the outside and with a structure of the exemplar which is free from isomorphic constraints.

A very similar hybrid strategy, which, however, involves placing the naturoid inside the organism, and therefore complies with some structural analogies with the exemplar, is used in the development of cardiac valves. In particular, for the valves designed by MASA (Mid-Atlantic Surgical Associates, New Jersey), researchers use tissues taken from ox or pig pericardium while the supporting structure is realized by polyester plastic materials. The advantage of using animal valves stems from the fact that, in such a way, one can avoid using anticoagulants to compensate for the organism's reactions against foreign materials.

Moreover, the same researchers adopt a completely artificial strategy in other cases, above all when the patient is younger and can bear the use of anticoagulants (or when the patient already uses them for other reasons). For this type of valve, the same ceramic material (which lasts a very long time) as that used for the covering of the space shuttle is employed; however, it requires the use of anticoagulants.

Finally, there is a long series of products which are midway between bioengineering and a technology aimed above all at restoring the body aesthetically—which we will deal with in a later chapter. These products include hand, wrist, shoulder and foot joints. Normally, at least in devices of greater bioengineering interest, silicone elastomers are used for their high endurance and their good biocompatibility, with titanium for the construction of the supporting parts.

Chapter 18
Artificial Limbs: History and Current Trends

It is said that the Rigveda, an ancient Indian poem, contains the first reference to a prosthesis. Written in Sanskrit between 3500 and 1800 B.C., it tells the story of a warrior, Vishpala, who, having lost a leg in battle, was fitted with a metal leg so that he could continue to fight. Likewise, the Celtic god New Haw was said to have four silver fingers.

In 500 B.C., Herodotus told of a man who brought a wooden foot. Wooden, bronze and leather prostheses have also been found in Roman ruins dating back to 300 B.C., and Pliny the Elder, in the first century B.C., wrote about a Roman general, Marco Sergio who, after the amputation of his hand, had a metal hand built to hold his shield and continue his military activity. In these cases, and in all similar ones that have surely accompanied military history, the devices were purely passive and were only intended to fill shortcomings of design or, at most, to act as a support to movement itself, as in the case of the artificial foot. Subsequently, in the Middle Ages, all kinds of prostheses made their appearance, both for military purposes and in order to hide deformities. During the Renaissance new impetus was given to the scientific study of nature renewing the efforts which had begun in ancient times (Table 18.1).

As we will see, progress in this area is very fast and amazing, but most of the obstacles encountered relate to our insufficient knowledge of biological systems we want to reproduce, and, more important, the lack of general effective and efficient strategies for integrating the natural and the artificial, that is, for enabling the biological system and the naturoid to collaborate without conflict, possibly to the benefit of the former.

It was not until the fifteenth century, in Germany, with the "Alt Ruppin hand", that the first artifact was developed which allowed, albeit in a rudimentary way, fingers to be placed in different configurations. In the sixteenth century a further innovation took shape. The famous hand of the knight Götz von Berlichingen was a device partially active, in the sense that, in addition to presenting a limb that looked like the original, it allowed him, by means of small buttons and levers, to

M. Negrotti, *The Reality of the Artificial*, Studies in Applied Philosophy, Epistemology and Rational Ethics 4, DOI: 10.1007/978-3-642-29679-6_18,

© Springer-Verlag Berlin Heidelberg 2012

Table 18.1 Five phases of artificial limb advancements

5000 B.C.–1500 A.D.	1504–1812	1812–1960	1960–1990	1990–2008
Aesthetical or simple support	Aesthetical and manual positioning	Based on a mechanical drive by residual natural muscles	Based on myoelectric signals transduction into mechanical forces	Based on digital processing of myoelectric or nervous signals

position the wrist and fingers through the desired movements, including the fundamental opposition of the forefinger with the thumb. In the same period the French physician, Ambroise Paré, introduced new artificial devices after amputation surgery and also devoted himself to the scientific design of implants.

In 1812, the surgical technician Peter Baliff first realized the possibility of giving the limb movement capability by exploiting the residual limb muscles of a naturally amputated person. In 1898, the Italian doctor Giuliano Vanghetti introduced a scientifically based kinetic prosthesis for exactly the same purpose, followed shortly after by the German physician Ferdinand Sauerbruch. The new technique was to link the physical muscle residues with the mechanical components of the prosthesis. The effectiveness of these devices, however, were reduced by the onset of inflammation and they were limited by the need to lengthen the muscles up to completely uncoordinated functionality of the prosthesis.

Since the end of World War II advances in prosthetics have been numerous and frequent. In 1946, the University of California at Berkeley introduced the technique of aspiration to optimize the connection between the device and the patient's body. In 1975, doctor Ysidro Martinez realized that, as in the other designs of naturoids, the obstinacy in wanting to faithfully reproduce the natural exemplar can lead to unnecessary delays and produce undesirable side effects. For this reason, when innovating an artificial leg, Martinez chose an alternative route by giving up the reproduction of the joints of the human limb but getting significant improvements in weight and in the acceleration and deceleration by the user.

In fact, the connection of the artificial leg to the body is a fundamental problem which was solved by adopting highly individualized strategies according to the age of the patient. The prostheses that worked best for young people were designed differently than the ones for older people. Devices that were valid only for those who had not been subjected to a bypass of the residual part of the limb were developed differently than devices for amputees who had a better chance of controlling the movement of the artificial devices and still differently for those who favored a balance of these different techniques depending on their body size.

However, from the 1960s on, the fascinating challenge became connecting the artificial limb to the muscular system, and then the nervous system became the focus of designers. This strategy replaced the pure use of the forces exerted directly by residual muscles with the withdrawal from the skin of the electrical signals that accompany muscle contraction and release. These signals, once amplified, were transduced into the corresponding forces exerted by actuators through electric motors or other components. Needless to say, in the decades immediately following, this innovation in transmitting functional nervous impulses became a refined process using microcontrollers that were increasingly miniaturized and better able to generate more flexible and complex movements proportionally to the intensity of the inputs and their combinations.

As you may have guessed, however, the *training* that is needed to make best use of such machines is not quick or easy. Despite the seemingly instant accessibility of the general model, which in theory allows the user to "drive" the device by means of the correspondence between voluntary movements and the movements of the prosthesis, what is achieved, although incomparably better than was possible in the past, meets only a part of the patient population demand. Many amputees end up, in fact, abandoning the use of even more advanced implants for their size, weight or appearance, are too similar to a robot, not to mention the rehabilitation of the mind that these innovations require. Something similar, as is known, happens in the case of devices that synthesize the human voice. They are clear, unambiguous and, therefore, effective and efficient, but, in practice, they are inherently incapable of reproducing the subtle modulations of the human voice.

Interestingly, even here, the transfiguration of the natural exemplar remains a limitation. In Karlsruhe, researchers developed Fluidhand, a device that is based on small cushions placed on the wrist that can move fingers when they are filled with air, thereby achieving a high degree of flexibility with less weight than in other devices.

One of the many problems that remain unresolved, however, is that of the limb's sensitivity to pressure, a fact crucial to the use of both a hand and a foot. The biofeedback available today, at least experimentally, allows for partial feeling of the pressure coming from the limb on the object with which the interaction takes place by means of electrodes connected to the skin of the patient through which signals flow that measure temperature. The same applies to the sensation of pressure exerted by the foot on the ground.

These solutions are of considerable importance because, among other things, they give rise to the "brain projection" that is to say, they enable patients to feel the entire artificial limb and not just the natural part which remains. However, researchers noted in 2003 that the adoption of a prosthesis adversely affects the generation of motor images by the patient. These images are generated by the same cortical areas of the brain that are active in the real movements of the limbs and anticipate its implementation.

The direct sampling of signals, using biosensors, from muscles or nerves of the patient is now taking the central role in bioengineering. In many ways, this area of study is restarting from scratch, setting completely new goals that, for example, give the patient an opportunity to directly control the actuators through the cerebro-mental expression of its intentions.

Devices of various kinds, using this simulation technology are being tested in several countries, including the US, Switzerland and Austria. Some include the implantation of electrodes into the muscles, others use solid-state electrodes around which nerves are allowed to grow. Other projects collect the brain signals from several sensors placed on the skin of the head and exploit the techniques already know from electroencephalograms (EEGs), looking, thanks to artificial intelligence software or artificial neural networks, to capture the correlations between the flow of signals and the commands the patient intends to send to the artificial limb.

The Brain Gate at Brown University in Rhode Island uses a device connected to the motor cortex of the patient, consisting of hundreds of microelectrodes a millimeter long and connected to a computer. "Luke Arm", a highly artificial robotic arm, created by the group of Dean Kamen, designed on behalf of the US Department of Defense, has as its objective the preparation of a limb which offers, through the use of several chips and motors, more degrees of freedom, 18, than the three normally available in current artificial arms, motivated by the fact that the natural arm possesses 22 degrees.

Once again, the project has faced the difficult problem of connecting the machine with the patient's body. The solution is provided, in this case, by the surgical technique of Todd Kuiken of the Rehabilitation Institute of Chicago. In it, the commands are given by the patient to "Luke" by activating intentionally the nerves that provide feeling to move the arm ('phantom limb'), putting in action the chest muscles where the nerves are connected, and thus activating, in the end, the actuators of "Luke." The patient, thereby, has the feeling of using his own limb. Biofeedback is also provided by a sub-system—called *tactor*—which generates vibrations on the skin, a technique also used to induce artificial effects of proprioception in fingers with the simple plastic-aesthetic purpose of reproduction. Alternatively, the patient can control "Luke" via joysticks operated by the foot.

Taken together, the current supply of artificial limbs, both experimental and commercial, thus presents some novelties of extraordinary interest. However, there remain difficulties which are due to the unavoidable heterogeneity of materials and the operating 'logic' of natural systems on the one hand and those of naturoids on the other. The connection with the biological organism, the flexibility of the implementation of the movements and the strategies for their control, are obviously three areas of intensive research.

The solutions that are appearing take the natural exemplars—eye, skin, arms, hands, fingers, legs, feet—as the only goals in principle, but without any ambition or ability to remain faithful to them until the end of the design and implementation. Even in our era, the true protagonist is still the general achieved level of technology and, therefore, the set of technologies actually available.

These technologies, and their "nature," impose constraints and chances from the combination of which the resulting naturoid can only arise and act in *sui generis* ways. The final performances of artificial limbs are sometimes more effective than those generated by natural ones. For example, it seems that Oscar Pistorius at the Beijing Olympics, thanks to his artificial legs, used only 25 % of the energy compared to a regular athlete; the strength, the endurance and even the feeling that an artificial hand can offer, in principle, goes beyond human capacity;

artificial skin may have a greater gripping power than natural and so on. In many other respects, the performance of the artificial is weak and only partially overlaps natural abilities. In the case of artificial limbs, flexibility, various orders of sensitivity and, possibly, their aesthetic configuration are still waiting for an optimal reproduction.

A recent area of research that, if followed by concrete achievements, will pave the way towards higher levels of diversity between the natural exemplars and the naturoids, is perhaps the most enlightening, despite appearances. A group led by the Faculty of Medicine, University of Florida, is trying to give the machine—the artificial limb—not only the ability to successfully execute the commands sent from the brain via appropriate connections, but also to learn and adapt to the behavior of the patient and to assist him with more efficiency. In essence, this means replacing the fixed algorithms that govern the microcomputer of the existing artificial limbs with algorithms that can be modified and optimized to converge towards the target accepting commands by the patient himself.

Future work seeks to combine artificial intelligence techniques with the requirements of an artificial organ, because the naturoids will incorporate a software that, instead of exploring the external context which is what a human being does when he decides to make a move, will explore the data coming from past similar experiences, allowing the prosthesis to anticipate the probable commands of the user. However, the exceptions, which make human actions so unpredictable, could be forced to undergo a standardized model within which the difference between the two intelligences, human and technological, would require some higher meta-level of coordination.

To conclude, we are today moving toward a partial convergence among projects coming from a number of different fields, namely artificial organs, robotics and artificial intelligence, that will share some aspects of the problems they work on, hoping to draw useful hints from each other. In fact, the final aim of the above three technologies is that of reproducing natural instances. But they also exhibit a great difference in terms of the benchmarks they should achieve. Robotics and artificial intelligence can work in a much more wide and free area, in the sense that their projects, though conceived to reproduce some human essential performances, often develop towards objectives that exceed the abilities of the natural exemplar without this transfiguration being considered as a failure. On the contrary, the projects in the field of artificial organs and limbs must fit both the natural physiologic performances and patient's expectations, which present themselves as a very demanding part of the environment within which the naturoid has to work.

Chapter 19
The Artificial Brain

Although the expression 'artificial brain', which was somewhat accepted in the 1950 and 1960s, is now obsolete, artificial intelligence research seems to be once again heading towards this objective after almost 50 years of attempts in which terms such as 'artificial mind', 'reasoning', 'understanding' and 'intelligence' have been used instead of 'artificial brain'.

By this we do not mean to assert that it is now time to go from pure metaphor to the concrete realization of an artificial brain. What we mean, on the contrary, is that research on artificial intelligence has led to the growing belief that the physical dimension—meaning all of the not merely informational conditions that characterize the work of the mind—not only cannot be overlooked, but is probably one and the same with the performance of intelligent activities in man. In short, even the most abstract intelligence is not born out of physical nothingness but is produced in relation to the continuous exchange of matter and energy with the organism's internal and external environments. In other words—and putting aside metaphysical considerations without denying their importance—the brain appears more and more to be what it actually is: an organ like any other organ, but, considered in its whole, much more complex.

Over the past 30 years, artificial intelligence has been called 'symbolic', since it maintained that it is possible to reproduce intelligence as an essential performance of the mind based on manipulations of symbols while completely overlooking structure and physical components, which are nevertheless very obvious and important in nature. In fact, researchers trying to understand brain functioning are taking many different paths. For instance, at the University of Southern California biomedical engineers try

> to treat the whole system [the brain] as what engineers call a 'black box.' In addition to trying to understand what goes on inside the box, we have tried to build a box of our own that behaves in exactly the same way, an electronic device that responds to any given input signal with precisely the same output signal that the natural neural system does (Berger 1995)

M. Negrotti, *The Reality of the Artificial*, Studies in Applied Philosophy, Epistemology and Rational Ethics 4, DOI: 10.1007/978-3-642-29679-6_19,
© Springer-Verlag Berlin Heidelberg 2012

Such an approach does not have any metaphysical understanding: artificial intelligence researchers do not think of the mind at all as metaphysicians would think of the soul, nor do they consider intelligence as one of its qualities. Rather, with an orientation which we can call functionalistic, they think, and thought, of the mind as the brain's software. Since software, as an informational structure, is transferable ad libitum on other physical supports, it can be inferred that even thought, understanding, reasoning and intelligence can be reproduced at will on any other supporting structure, for example the computer.

Therefore, the foundation of this approach lies in the above mentioned principle of functional equivalence. Authors such as P.N. Johnson-Laird and Z. Pylyshyn have thoroughly explored this subject, the true cornerstone of the whole artificial intelligence case or, at least, what J. Searle defined as 'strong' artificial intelligence (Searle 1984), according to which thought ability actually takes place in the computer.

In order to further clarify the principle in question, it is worth reading carefully the following quotation from the physicist Craik, quoted by Johnson-Laird in 1988:

> When we speak of model we thus mean every physical or chemical system whose relationships are structured as in the process which is imitated, that is, in which there is an equivalent structure of relationships. By 'structure of relationships' I do not mean a mysterious non-physical entity which accompanies the model, but the fact that there is an operating physical model which functions in the same way as the process which is parallel to it. (Johnson-Laird 1988)

Now, if we consider the actual results and products which artificial intelligence has produced throughout these years, we realize how groundless and useless for the discipline's progress those claims were.

The most efficient computer programs, inspired by artificial intelligence models, are certainly expert systems, software capable of reproducing human reasoning—of an expert in a specific field such as medicine or law, geology—for example, of a diagnostic type, by operating on two elements: a 'knowledge base' and an 'inferential engine'. Whereas in the former the knowledge of human experts is coded, which makes it available through meticulous and complex interviews, the latter consists of an algorithm capable of performing deductions, inductions and probability calculi.

In order to illustrate these concepts let us consider a general example. If the concept that "illness M is identified by symptoms A, B, C, D" is inputted into the knowledge base, the expert system, which is required to provide a diagnosis for someone with symptoms B and D, will ask if symptoms A and C are also present and, if this is the case, it will assume illness M. The expert system could present the same diagnosis even without symptom C, if the human expert who instructed the computer gave it less or additional importance compared to A, B and D.

Clearly, such a system can be presented in much more complex terms according to the amount of knowledge that is at the expert system's disposal and the various methods of analysis and prediction it is equipped with.

Nevertheless, a few objections can be made that, of course, concern not the usefulness of these programs, but rather the pretense that an expert system reproduces the mental activity of a human expert. Deductive activity, based on an appropriate representation of knowledge, is certainly a performance of the human mind, but why should it be the 'essential' one? The fact is that, as M. Minsky and R. Schank—two artificial intelligence pioneers—have observed on many occasions, every model, in working on human intelligence, offers something useful but, nevertheless, something which escapes its attention, sooner or later, determines its limits and inadequacy.

Put in our lexicon, we must acknowledge that, when designing a model of the mind or even only of intelligence, every observation level offers some advantage, to the detriment of other aspects which invariably prove to be crucial on various occasions. Thus, as regards intelligence, the limits of expert systems lie in their inability to receive, and thus reproduce, intuitive, synthetic or very personal knowledge, which, however, constitutes a good share of the intelligence that a human expert resorts to. This difficulty cannot be attributed to a particular defect in programming the machine: rather, it is due to the fact that the human experts themselves are not always able to explicitly state the line of reasoning they employed in developing their diagnosis, prediction, and so on. In these circumstances we speak of 'tacit knowledge' (Polanyi 1966), which includes all the mental abilities that, although they are real and produce real effects, cannot be completely expressed even by the person who possesses them. Other strategies, such as learning by example—which, to a certain extent, should reproduce the learning that takes place when a good assistant works next to an expert—have not improved the situation very much.

This sort of limitation can be attributed to the fact that computers are supports and processors of information, and not knowledge. Production of knowledge implies information processing but information processing, as such, does not imply *ipso facto* production of knowledge. More importantly, the rules according to which we process information—which can easily be transferred to the machine—have nothing in common with the ways we think. Besides, the informational observation level is important in the mental world, but it is not easily isolatable from the whole of events that characterize it.

After all, even a simple electronic calculator processes data by developing mathematical calculi, and the deduction is nothing but a calculation. Once it is isolated, let us say almost eradicated, from all the dimensions of mental life—which are moreover unknown in their entirety—the ability to develop mathematical or logical calculi is presented as a very relevant but arbitrary essential performance: it is very useful where the problems to be solved are compatible with it and completely insignificant where the problems are of a different nature, such as those related to several forms of common sense, intuition, criticism, suspicion, and, most of all, problem finding.

Since artificial intelligence has never been able to seriously consider the idea of reproducing the structure of the natural exemplar (the brain), it is clear that its falling back on the principle of functional equivalence is adopted in order to capture some essential performances in themselves. In this sense, artificial intelligence has been successful, just as pure and simple automatic calculus has been successful. The different hardware and software support—or, in our lexicon, the different materials and procedures—adopted to realize intelligent devices has nothing to do with what Craik calls 'model', in which

> … the relationships are structured in the same way as the process which is imitated … (Johnson-Laird, cit.)

Such is the case for the simple reason that the mind is a largely unexplored exemplar and, moreover, it is very strongly open to social representations which, in turn, are influenced by current doctrines, extra-scientific paradigms, and so on. Therefore, even if we would have some reasons to think that something like a mind 'exists' as a natural exemplar—and not as an essential performance of the only natural exemplar we can empirically observe, namely the brain—it is a matter of fact that it is very difficult to assign reliable boundaries to such an exemplar.

In contrast, it is a plausible idea that, as in many other cases of naturoids objects, processes or machines, even in artificial intelligence's best results we note the definite priority given to some essential performance. Even if it is only of an informational nature, it is an aim to achieve, and in fact pursued in various aspects. Indeed, it would be difficult to deny that an expert system is intelligent, at least within the limits of intelligence understood as the ability to use information rigorously. Hence, we can claim that artificial intelligence produces real intelligence but also 'authentically artificial' and, so to speak, closed intelligence.

Towards the end of the 1980s, in the wake of the disappointment regarding the lack of success of 'strong' artificial intelligence, a research tradition which dates back to the 1950s—but which has its foundations in the logical works of McCulloch and Pitts in 1943—came back to the fore. This is the tradition of the so-called artificial neural networks. Even in this case, reference to the exemplar—the brain as a structure of physical connections between neurons—is little more than evocative, even though research on neural networks, unlike symbolic artificial intelligence, can rely on some scientifically established knowledge. Basically, as we have said before, the neural networks reproduce the model we have of the connections between the neurons which characterize the brain, but, obviously, instead of the structures, tissues and biochemical signals of this organ, the components adopted are electronic information units coming from mere electrical signals.

According to the general definition advanced in 1988 by DARPA (Defense Advanced Research Projects Agency), a neural network

> … is a system composed of many simple processing elements operating in parallel whose function is determined by network structure, connection strengths, and the processing performed at computing elements or nodes" (*Armed Forces Communications and Electronics Association*, 1988)

According to Haykin,

> A neural network is a massively parallel distributed processor that has a natural propensity for storing experiential knowledge and making it available for use. It resembles the brain in two respects: 1. Knowledge is acquired by the network through a learning process. 2. Interneuron connection strengths known as synaptic weights are used to store the knowledge (Haykin 1994)

Since even in the human brain learning seems to imply strengthening the synaptic relationships between neurons, we can say that the neural networks reproduce the exemplar's structure, namely the architecture of the brain, at least at this 'network-based' observation level.

The usefulness of neural networks lies in their capability to solve problems for which there are no symbolic algorithms or these algorithms are too complex to design. As paradoxical as it may seem, it is actually true that the symbolic or mathematical description of the process through which we quickly recognize our car in a crowded parking lot without any calculations is much more complex than the description and calculation of an astronomical event. Neural networks thus try to simulate the identification process by imitating the brain's automatism—or, better yet, its self-organization—rather than by employing an approach which calls for a formal and conscious analysis.

The idea is that the presentation of a certain data configuration to the network—the numbers which describe a certain computerized picture, for example—strengthens certain connections between neurons rather than others, and therefore similar configurations strengthen the same region of connections more and more. Hence, an appropriately trained neural network is able to distinguish a specific configuration (for example a fingerprint) from others or from the background. It should be noted that the network is governed by mathematical criteria of calculus (of the relationships), but it does not follow any algorithm, that is, it does not execute any program, as does symbolic artificial intelligence.

It is evident that these two currents of thought on which artificial intelligence is based privilege two groups of essential performance which are different from one another, yet present in our brain at the same time. Furthermore, it does not seem at all easy, as we have maintained with regard to all naturoids, to find a way to get a neural network and a symbolic artificial intelligence software to work together, without having to face the problem of finding and describing a superior observation level and model and, possibly, the natural exemplar it refers to. The latter should include the first two not in an arbitrary manner, but in a way which is as similar as possible to what happens in the natural brain—something that, at this point, could be called 'mind'—in which performances of recognition and formal informational or even symbolic performances are certainly present and coordinated amongst themselves.

A recent research development in artificial intelligence foresees the opening of this discipline to the social sciences. It is a new approach which, considering the fact that human intelligence is developed in the environment, regards the social relationships between software or hardware 'intelligent agents' as models of study and design. Some of the main topics regarding this approach, for research on components of social relationships, include: credibility, imitation, cultural adjustment and the coordination between the internal or external dynamics of the agent.

The model is therefore made up of society, in a simplified or implicit version. However, the essential performance varies from project to project, leaving open, as usual, the problem of their coordination in an integrated framework. Such a framework, could only be found in some general theory of social relationships, namely in a field in which sociologists themselves have been working for at least a century, without accomplishing anything definitive.

In any case, man's role itself becomes odd at this point, both as a designer and as a user, since, at the moment of designing and using similar intelligent agents, he will inevitably become related to these agents, setting off unpredictable phenomena of mutual adjustment. This is a topic which concerns many types of intelligent naturoids, in which man himself becomes a part of the process he designs (*human in the loop*) by trying to give the machine the ability to adapt to human design.

In conclusion, all the cases of artificial intelligence which we have examined, are evidence, on the one hand, of the ambition and necessity to find useful exemplars in nature—in man, in society—in order to design intelligent machines or processes, and, on the other hand, of the inevitable transfiguration of those exemplars and their performances by the resulting naturoids.

For example, an expert system not only reproduces some of a human expert's explicit abilities, but it can do so with extreme speed, accuracy and without ever forgetting those abilities, even in environmental situations in which man would find himself in serious difficulty. Furthermore, an expert system can draw on a wider range of expertise, making use of the skills of different human experts. Such a system can therefore operate as if it had more than one personality, precisely because it has no personality and its way of reasoning, unlike man's, can be varied *ad libitum.*

Neural networks, though the issue also concerns the 'sociological' perspective of artificial intelligence on a different level—often lead to completely new diagnostic or recognition behaviors, be they erroneous or simply 'mysterious'. Hence, they offer the possibility of speculating on alternative ways of interpreting and interacting with reality, on the basis of which new models and styles are already being developed in the fields of play, art and the design of machines destined for various non standard functions.

Likewise, ALife programs, although they were born out of the desire to capture the essence of natural life and evolution through algorithms, end up providing researchers with models of life "... as it could be" (to use a famous expression by Chris Langton)—as well as models of life for entertainment products and other kinds of products. In conclusion, in this field of naturoids as well, even though the objective of the reproduction of the exemplars and their essential performance is only partially achieved, 'something happens' and it should not taken for granted that this something is less interesting or important than the unsuccessful reproduction objective.

Chapter 20
Prostheses, Replacements and Surrogates

Findings of prostheses have been brought to light since the fifth Egyptian Dynasty. Recently, on the border between Iran and Afghanistan, an artificial eyeball was found dating to 5,000 years ago and constructed of lightweight materials derived from bitumen paste which was carefully finished with the design center of the iris and an effective gold 'ray of light'.

The category of artificial objects, processes and machines of a purely aesthetic kind—and therefore without any pretense of reproducing the exemplar as far as structure or functionality are concerned—is as ancient as it is vast. It includes activities that range from developing substitutive prostheses of organs or body parts to the reproduction of natural components of landscape, reproducing nature's scents to imitating or 'false' products, for commercial purposes, which get their inspiration from preexisting technological products. For instance, as far as leather is concerned, it

> ... was first created in the US in 1963 and was later introduced to Japan, marking dynamic growth. Up until the 1970s, artificial leather was considered an alternative to real leather, and in the 1980s it was considered healthier and cleaner than the real thing. Then in the 1990s, when the necessity of conducting environmental countermeasures on a global scale became apparent, artificial leather manufacturers strove to develop products that would place less strain on the environment (Teijin 2000)

Although there is no room in this work to discuss it at length (a more complete examination can be found in App. B and in the References at the end of the book), it is useful to keep in mind that art itself—just think of figurative painting and sculpture—has a clear ancestral, reproductive role. Indeed, we cannot deny that all artists intend to reproduce objects and processes from the outside world or their own mental states as natural facts, interpreting them according to their own poetics, using materials and procedures which are different from those that we find in an event in nature or, presumably, in one's mind.

The only basic difference between the reproduction of a figure in a painting and an anthropomorphic robot in the technological field stems from the markedly

M. Negrotti, *The Reality of the Artificial*, Studies in Applied Philosophy, Epistemology and Rational Ethics 4, DOI: 10.1007/978-3-642-29679-6_20,
© Springer-Verlag Berlin Heidelberg 2012

different views of the artificial and the natural regarding transfiguration. While for a robot designer, the inevitable transfigurations of his product with regard to the exemplar and its performances are often a source of disappointment and frustration or even of astonishment over something unexpected, the main aim of the artist's work is precisely that transfiguration of the exemplar and its performances. Indeed, even the most 'realistic' artist reproduces the world with a deliberate interpretive intention and not as a mechanical copier. Moreover, this sort of mechanical copying is impossible to imagine for the reasons which apply to all artificialists. The beauty of a painting lies in the persuasiveness, power and elegance of its projection of a reality and not in its reproduction of reality as such or, better put, a reality as it commonly appears and is described by human beings in a given historical period.

The aesthetic level on which reconstructions and various types of gadgets are found is naturally different and, here, the designer's ideal ambition is to reproduce reality precisely 'as it is', that is, as it is perceived and represented collectively. As we know, various parts or organs of the human body have always been the subject of these kinds of reproduction attempts, not only for personal reasons, but also because in many civilizations and historical periods mutilation was feared even more than death, and it was often full of negative meanings.

Nevertheless, as a rule, this area of naturoids, which is also almost always oriented towards the maximization of the aesthetic effect, now has a very particular relationship with bioengineers' concrete-analytical reproduction attempts.

The human eye is perhaps one of the most interesting examples. The traditional prostheses which we know about—invented in 1938 by an American doctor, Fritz W. Jardon—are often designed very well, but limited since their eye-ball is motionless while the normal eye-ball is in constant movement.

But there is now a new kind of artificial eye, designed by Bio-Vascular, Inc. Thanks to a special biomaterial, hydroxyapatite, muscle tissues grow on the back of the prosthesis and these tissues, with the aid of additional mechanical devices, allow the artificial eye to move simultaneously with the natural eye. In short, it is a case of double illusion. The organism, moves the artificial eye-ball with its muscles 'thinking' that it is the natural one, while at the same time, an outside observer is led to believe that the person with the prosthesis has completely normal eyes.

Obviously, this kind of artificial eye does not become an actual part of the organism and various precautions and maintenance activities are necessary. At any rate, the majority of these prostheses last 8–10 years because the form of the eye socket changes over time, hence the device no longer fits properly.

Today, hair, above all, constitutes an artificialization objective which is pursued very actively. In light of the immunity problems associated with natural hair transplants, artificial hair technology has developed extensively. Above all, it makes use of synthetic materials such as nylon, PET (a polyethylene) and modified acrylic substances. The goal is to reproduce the natural exemplar in its exterior appearance—perhaps the essential performance selected almost all the time—but there is also another consideration, namely the flexibility of natural hair.

As is so often the case, the main problem stems from the relationship with the human body. Considerable efforts are being made here as well, including the use of silver to cover the hairs, thus reducing the chance of infection. Nevertheless, all artificial hair is eventually rejected as an extraneous body. In light of such an ever-present and intrinsic danger, in 1984 the American FDA declared the implant of such devices illegal, while it is still legal in Japan, Mexico and Europe, though surrounded by numerous medical controversies.

Proceeding to naturoids which do not assume human body parts as exemplars but rather natural objects or processes, in the food sector we find considerable efforts to 'surrogate' natural substances or products. Among the thousands of examples which we all know, just think of the so-called sweeteners or artificial sugars, whose growth was given further impetus after reports were issued on the negative effects of natural sugar. Although such reports were rather exaggerated giving rise to one of the many collective missions full of hysteria, the industry of artificial sweeteners has generated various products, none of which, however, has come on the scene without producing unpleasant or harmful effects owing to the unavoidable principle of inheritance. Saccharine, for example, gives a pleasantly sweet taste, but then leaves a bad aftertaste on the palate; aspartame, in turn, does not have a bad aftertaste, but it seems that aspartame/phenylalanine's metabolites (one of its components) include several toxins, such as methyl alcohol, which can be dangerous even in small quantities.

However, some companies, such as California's Wholesale Nutrition, maintain that artificial honey, which is made from ascorbic acid and various essences, is cleaner and less harmful than natural honey, since it does not contain adulterants, botulin spores, drugs, pesticides, allergenic factors, bee residue or waste, or other insoluble elements that come from beehives. In short, in giving the recipe for the product, a Wholesale Nutrition expert maintains

> … we offer our kit to bring our clients' and the world's attention to the insidious nature of all sugars. Without our kit, nearly all of us would continue to ingest about 45 kilos of saccharose, glucose and fructose a year with great harm for our health (Wholesale Nutrition 1996)

But efforts to cope with nature cover the whole spectrum of flavor, and, as stated by a manufacturer,

> Through careful analysis of key flavor ingredients, Flavor Concepts can help develop natural and artificial food flavorings rivaled only by Mother nature herself—or we can create new flavors, combinations and tastes she never even tried (Flavor Concepts 1998)

Differences between the taste of natural and naturoids are sometimes intentionally reviewed by newspapers, as is the case for the beer. According to the *San Jose Mercury-News*, artificial beer

> … has a refreshing taste, though a bit sweet, and is best when mixed with extremely cold water. Its taste is remarkably similar to beers produced by micro-breweries. It is a quick source of liquid carbohydrates, and it is easy and light to pack and mix … In very cold water the mix clumps up unless you add water slowly and stir constantly; although it does not compare to a fine lager, it suffices quite nicely when your taste buds crave a cold one in

the backcountry and you do not fancy carrying a six-pack. The manufacturer mentions one can add clear grain alcohol or vodka to achieve an alcoholic beer (*San Jose Mercury-News* 1997)

The claims of manufacturers, but also the many popular beliefs that support this or that feeding fashion, should be carefully investigated, because, even regarding seemingly simple use of naturoids, something new will always appear, as in artificial rearing (Kaneko et al. 1998).

In other special circumstances where natural phenomena are well understood in their basic composition, and their architecture is easily reproducible, man is able to get very close to the exemplar and its essential performances. Diamonds are a case in point:

Most of pure diamond's fundamental properties are retained in artificial diamonds. For example, artificial diamonds have extreme hardness, broad transparency, high thermal conductivity and high electrical resistivity (Weintraub 1998)

Likewise, various movements and manufacturers say "no to artificial tastes", "no to artificial colors", "no to artificial sweeteners", "no to artificial preservatives", and fight their battle for a return to nature. Whoever has any doubts about the distinction between natural things and naturoids should find a lot of material to reflect on in this almost ideological open campaign. Although all the products we ingest are actually made from elements, compounds and substances that exist in nature, the distinction between natural and artificial is apparent to everyone. Such a distinction is easy to make since the combination—good or bad—which nature gives to the elements is by definition different from the one man creates when he attempts to reproduce it as such and, obviously, when he actually attempts to improve it.

Chapter 21
Artificial Environments and Landscapes

Among the objects of nature which are today most often assumed as exemplars, we can find elements of the landscape or climate, such as rocks, snow, grass, rain, islands, caves, mountains, ponds, lakes and many others.

Artificial rain—an undertaking which assumes a process, rather than a structure, as its exemplar, just like artificial insemination—has a rather long history, at least in Kansas where in 1890 a certain Melbourne ("The Rain Wizard") shot mysterious gases from his roof in order to stimulate the meteorological phenomenon in question. As far as we know, Melbourne, did not possess great secrets and his practices were not much different, apart from references to something 'scientific', from the magical practices of rain propitiators. However, in spite of his failures, a few private artificial rain companies came into being, each one claiming to have obtained from Melbourne the technology he used.

In Kansas, the Western Kansas Weather Modification Program has been in operation since 1975, and one of its goals, besides artificial procedures to prevent the formation of hail, is to increase annual rainfall. There is even a local museum on this topic where one can examine various apparatus, including a device for the insemination of clouds. The rules of "cloud seeding" were discovered in Schenectady, New York, by Irving Langmuir, Nobel laureate in chemistry in 1932 and director of General Electric's research laboratories.

The rainbow is one of the most spectacular natural phenomena humans have tried to reproduce. Today, several techniques are available, ranging from very simple means you might find in a physics classroom to very sophisticated technologies. Among the latter are the Hitachi diffraction gratings at the Okazaki National Research Institute, National Institute of Basic Biology. According to Hitachi

> The spectrograph has successfully realized the world's largest artificial rainbow whose intensity is 20 times that of sunlight energy right above the equator! Hitachi plane diffraction gratings consisting of varied space grooves have also been adopted in the spectrometer of the extreme ultraviolet explorer launched by NASA (Hitachi 2001)

M. Negrotti, *The Reality of the Artificial*, Studies in Applied Philosophy, Epistemology and Rational Ethics 4, DOI: 10.1007/978-3-642-29679-6_21,
© Springer-Verlag Berlin Heidelberg 2012

Unlike rain and rainbows, artificial snow has a pragmatic purpose at times and an aesthetic purpose at other times. As we know, the main pragmatic function of artificial snow is to make the sport of skiing possible. The undertaking adopts the structure of snow as an exemplar. This snow is produced according to different principles than those of natural snow, using water atomizers and compressors. The result is almost always satisfactory, although, according to some northern European skiers, artificial snow produced in the southern parts of the continent, unlike the one available in northern European countries, often tends to be 'pure ice which is used out of despair'. In other cases, focusing more on the functional equivalence principle, the sporting aim is pursued by choosing a more restricted essential performance—the viscosity necessary for skiing—rather than of the production of snow: this is the case for artificial ski slopes made using materials such as polyethylene, which lasts longer and guarantees an ideal flowability with limited friction. Furthermore, it seems that in presence of artificial snow

> The vegetation reacts with changes in species composition and a decrease in biodiversity … artificial snowing induces new impacts to the alpine environment. Snowing increases the input of water and ions to ski pistes, which can have a fertilising effect and hence change the plant species composition … snow additives, made of potentially phytopathogenic bacteria, are used for snow production. They enhance ice crystal formation due to their ice nucleation activity (Rixen et al. 2003)

Conversely, the aesthetic purpose calls for a type of artificial snow which would be difficult to ski on, since the selected essential performance only concerns its appearance. Once again, the same exemplar lends itself to two or more different versions of artificialization, without any possible relationship between them. In the aesthetic case, it is a fibrous product, readily available on the market, which can be deposited on the ground to simulate the presence of snow and, once deposited, as the producer of one of these substances affirms, you can create various effects, from footprints to tire prints. However, there are considerable side effects lying in ambush, in particular, the danger that the fibers can drift in the air and cause irritation if they are inhaled.

Definitely, even the most seemingly simple naturoid, be it an object or a process, cannot hope to reproduce completely its natural exemplar and its essential performance, when, after being conceptually and concretely isolated through a model from its natural context, it is placed and put to work in the real environment. Thus, it is not surprising that even the artificial drying of plants does not escape this rule, and it has been found that:

> Artificially dried, immature corn was more brittle than air-dried corn. Lyophilized samples contained more sugar and niacin, but less starch than air-dried corn (Gausman et al. 1952)

The same applies to the process of artificial defoliation, since, according to a study conducted in Cruger, Mississippi.

> On average, undefoliated trees were significantly taller than defoliated trees. Trees defoliated at the rates of 50 percent and 75 percent exhibited the greatest decrease in growth. Height growth for trees defoliated at these levels was reduced by 30 and 31 percent, respectively (Tucker et al. 2004)

Another field worth mentioning, which assumes a process rather than an object as its exemplar, is so-called 'artificial weathering'. It consists, for example, in monitoring certain organic coatings of various kinds of natural objects by subjecting them to regular and accelerated conditions using ultraviolet rays. Normally, as the Swiss Federal Laboratories for Materials Testing and Research attest, there are not any great similarities between weathering caused by artificial methodologies and weathering caused by nature, due to the numerous variables included and the presence of very complicated decomposition processes.

We should also mention the field of 'special effects' in films. This field, called fiction, has always made ample use of devices which tend to give the spectator the most realistic impression of all kinds of situations and events. Specialized businesses, dedicated to meteorological events, like the ones produced by American Sturm's Special Effects International, are able to reproduce winter the whole year long, regardless of atmospheric conditions or climate. Moreover, from the lightest shower to the strongest storm, from the lightest wind to cyclones, specialists can simulate the climatic conditions of any season. Of course, in order to appreciate these reproductions, one has to place oneself at a very special observation level, in order to avoid 'breaking the rules' which allow the reproduction to appear as a natural phenomenon and, therefore, to deceive.

Another ambivalent analysis can be made of artificial grass, which is useful in some sports but also made for decorative purposes. Its early version dates back to the 1950s when the Ford Foundation promoted the design of a synthetic fiber that has progressively developed, acquiring several new properties. The TenCate Nicolon product, called Thiolon Grass, uses particularly steady polymers with a high thermal conductivity and therefore a great ability to absorb heat. Hence, according to the producer, artificial grass made in this way has characteristics which are actually better than those present in natural grass. Indeed, it can be used regardless of the temperature of the place where it is installed; it is produced in several varieties according to the sport; it cushions falls better than natural grass; it is resistant to ultraviolet rays and, of course, it does not require fertilizers or herbicides. No undesirable effects are mentioned, though the producer says he is sensitive to environmental problems and he makes sure that the pigments used are well tolerated by the environment.

With regard to artificial herbicides or fertilizers, there is also the long-standing controversy, concerning possible health risks associated with their use. In principle, we cannot claim that something is harmful only because it is artificial. Whether natural or artificial objects or processes are harmful or advantageous is determined by their structure, that is, by the above-mentioned combinations and recombinations of the same natural elements and not because they are the work of nature or man. The ancient fear of the 'devilries' carried out by technicians—or scientists—today is still a difficult enemy to defeat. Overcoming this fear, on the other hand, can at times leave too much room for enthusiasm for a reproduction or an improvement of the natural world, by means of technology, without any serious control.

Natural landscapes, in turn, have been altered by man for purposes which often have a prevailing conventional technological purpose—a city, a road, a bridge—but which often try to locally reproduce natural configurations assumed as exemplars for economic, touristic or military purposes. This is the case of the crannogs, or artificial islands, built in Scotland probably during the Neolithic Age and inhabited up until the sixteenth century. Several English university studies are now trying to clarify how geological and environmental conditions influenced the construction of these islands and what relationships their inhabitants had with their natural surroundings. Likewise, in Japan, the Kasai artificial beaches were built to the extreme north of the Bay of Tokyo and, here as well, research is under way to establish what chemical and dynamic changes are affecting the bottom of the sea following this human intervention.

On a purely aesthetic level, but with all the social and cultural functions associated with it, the technology, or art, of gardening deserves a separate comment. This technology, which since ancient times has created real, local artificial landscapes, involves relationships with the natural context that are often essential.

With regard to gardening, just consider a passage by Edgar Allan Poe from his *The Landscape Garden*, written in 1850, in which his interlocutor, Ellison, quotes a writer who expresses himself in this way:

> "There are, properly," he writes, "but two styles of landscape-gardening, the natural and the artificial. One seeks to recall the original beauty of the country, by adapting its means to the surrounding scenery; cultivating trees in harmony with the hills or plain of the neighboring land; detecting and bringing into practice those nice relations of size, proportion and color which, hid from the common observer, are revealed everywhere to the experienced student of nature. The result of the natural style of gardening, is seen rather in the absence of all defects and incongruities—in the prevalence of a beautiful harmony and order, than in the creation of any special wonders or miracles. The artificial style has as many varieties as there are different tastes to gratify. It has a certain general relation to the various styles of building. There are the stately avenues and retirements of Versailles; Italian terraces; and a various mixed old English style, which bears some relation to the domestic Gothic or English Elizabethan architecture. Whatever may be said against the abuses of the artificial landscape-gardening, a mixture of pure art in a garden scene, adds to it a great beauty. This is partly pleasing to the eye, by the show of order and design, and partly moral. A terrace, with an old moss-covered balustrade, calls up at once to the eye, the fair forms that have passed there in other days" (Poe 1856)

Man-made landscapes can therefore be more beautiful than natural landscapes, as long as the relationships between the two contexts, as in every other case of naturoids, are carefully designed, as the best architecture of the past centuries has taught us.

This kind of problem was faced by the architects and various specialists who, in Malibu, on the California coast, executed the well-known reconstruction of a Roman villa (the Villa dei Papiri of Herculaneum, buried by the Vesuvius eruption) by Paul Getty, a rather meticulous undertaking accomplished with almost all the same materials as the exemplar, thus giving rise to a pseudo-replica, which, nevertheless, requires a great deal of maintenance since it is situated a few steps away from the Pacific Ocean. Moreover, the same difficulties have to be faced not only by the people who take care of zoos, which we mentioned in a previous chapter, but also by those who try to reproduce natural environments in which an animal or vegetable species, such as aquaria, pods, nests, reserves, and so on, can survive. Furthermore, the already quoted case of the Japanese *domes*—with their very demanding need for maintenance—may be the most impressive because the larger a naturoid of this kind is, the wider is its simulation of nature's heterogeneous actions. In fact, including an artificial landscape that takes as exemplar a natural one, which may reside thousands of miles away in a host's landscape, results in the insertion of a 'foreign body' into an existing body, deliberately inviting the latter to attack the former.

Chapter 22
Virtual Reality

Deception, in the broader sense intended here, illusion, through a whole series of side effects, and also the potential utility of the artificial, converge, in the end, in Virtual Reality technology. This technology consists of devices which generate three-dimensional moving pictures on a stereoscopic monitor applied in front of the eyes in a special helmet. It is an artificial environment in which it is possible to interact, for example, by virtually moving in a room or virtually exploring the human body from the inside. Although this technology has aroused the usual controversy between enthusiastic supporters and detractors who are afraid of the possible 'loss of a sense of identity and reality', some of the most interesting applications are once again in the field of medicine. Indeed, it is possible, by means of virtual reality machines, to visit a patient or operate on him (telesurgery) even though he is far away from the doctor. In 1995, the demonstration presented by the DARPA (Defense Advanced Research Projects Agency) biomedical program was oriented in this direction; this program consisted of carrying out a surgical procedure with a robot which acted on the command of doctors who 'operated' using a monitor many kilometers away from the patient. Thus, we have a chain of naturoids (the viewing of the patient's body by means of telecameras and computers, the robot's arms and hands which intervene on the physical reality of the patient) whose coordination presents several difficulties, including the transmission speed of the signals in both directions, which is most important.

The most curious aspect—which calls to mind man's ability to adapt to the machines he uses—consists in the possibility that the surgeon, adapting to the delay which is generated between the moment in which he 'acts' on the monitor and the moment in which the robot acts on the patient, may become confused when he takes off his helmet. Moreover, according to experiments conducted by the British Defence Research Agency, virtual reality devices can generate several psycho-physical problems: out of 146 adults in perfect health who used the helmet for 20 min, 89 suffered temporary nausea, dizziness or eyesight trouble and eight did not even succeed in finishing the experiment (Langreth 1994).

M. Negrotti, *The Reality of the Artificial*, Studies in Applied Philosophy, Epistemology and Rational Ethics 4, DOI: 10.1007/978-3-642-29679-6_22,
© Springer-Verlag Berlin Heidelberg 2012

At any rate, for providing an effective aid to surgery, many believe that the essential performance chosen up to this point, namely the movement and thus the action of the surgeon's hands, is not sufficient. In fact, these critics believe that it is necessary to introduce devices into the system which are able to reproduce other typical elements of a surgical operation, such as odors and physical sensations which are associated with the forces involved. Additional research is dealing with these problems as well, and even in this field we can foresee the problem of the model on the basis of which the various performances will be coordinated. According to a report by the American National Institute of Standards and Technology

> The manipulation of instruments by the surgeon or assistants can be direct or via virtual environments. In the latter case, a robot reproduces the movements of humans using virtual instruments. The precision of the operation may be augmented by data/images superimposed on the virtual patient. In this manner the surgeon's abilities are enhanced (Moline 2001)

But

> the sense of smell in virtual environment systems has been largely ignored. Both Krueger and Keller are developing odor-sensing systems. Smells are extremely important. Not only do they help distinguish specific substances, but also they give a sense of reality to a situation. The absence of odor is a serious limitation of current telepresence and training systems. Another major research problem relates to overlaying ultrasound images on live video that is then viewed in a head-mounted device application. The research issue to be addressed is the alignment of images in real-time (ibidem)

Anyway, according to the report,

> Today simulations trade off less realism for more real-time interactivity because of limited computing power, but the future holds promise of a virtual cadaver nearly indistinguishable from a real person (ibidem)

As far as the problem of the odors is concerned, it should be stressed that recent research has provided some steps forward. In fact, technology is now able to reproduce odors and scents, allowing them to be embedded, at least in an experimental way, within any delivery system that needs their action.

At the time of this writing, there were no known medical simulators that incorporated computerized scent technologies into their framework. Anyway, some models have been proposed that, for instance, exploit an artificial nose which

> … identifies odors in remote surgical environments and then electronically transmits the electronic signatures for those odors over a computer network to a separate location, where the odors area recreated for the coaching surgeon's benefit. (Kumar and Marescaux 2008)

It is a matter of a rather complicated solution which surely will need further improvements in the future, particularly regarding the real-time performance and overall coordination of all the phases of the involved processes.

Virtuality is not a prerogative of advanced technology, since our imagination is its first creator, and this is why man has always created objects or machines capable of stimulating it, with or without a computer.

Environments and landscapes can also be observed through a window, as the American company BioBrite, from Maryland, has intuited.

> People prefer rooms with windows, and for a good reason. Specific research has demonstrated that windows can raise people's morale and even increase productivity. Now it is possible to take advantage of Window-Lite, a pleasant and economic solution for offices, apartments, hospitals, workplaces or other places which are not equipped with windows (BioBrite 2000)

It is a sort of illuminated picture which reproduces the structure of a window and allows one to admire landscapes which range from classical Hawaii to tropical beaches, while creating the sensation of space and, according to the producers, reducing the symptoms of claustrophobia. All things considered, we find ourselves before a technological remake of the quadraturism and the *trompe l'oeil* which date back to the sixteenth century, though with a more direct, and maybe na, intention of obtaining realistic effects thanks to the technology available today. Technology allows the same above-mentioned company to offer a Dawn Sunrise Clock simulator piloted by an electronic watch and other devices, which acts as an alarm clock, but

> … by reversing the negative effects of a gloomy winter day, minimizing various physiological problems associated with daily rhythms and helping people to wake up in a natural way (ibidem)

Chapter 23
Conclusions

Conventional technology and the technology of naturoids, respectively, generate machines and processes as non-natural realities. While conventional technology aims right from the beginning to give rise to things which do not exist in nature, the technology of naturoids, though it wishes to generate things inspired by natural exemplars, cannot avoid transfiguring them to some degree.

A world made up exclusively of conventional technological objects and machines would be a world in which two realities, the natural and the technological, would be, as in fact they are, very distinguishable from each other, just as a city can be distinguished from the surrounding countryside or, better yet, a car 'cemetery' from the landscape in which it is situated, or a watch or bracelet from a human being's wrist or a ball-point pen from his fingers.

Conversely, naturoids, at least ideally, aim to create realities which at certain observation levels should not be distinguishable from the natural context, whether it be the physical environment, the human body or any other natural environment. Naturoids, in fact, stem from an ancient desire not only to control nature but to reproduce it using strategies that differ from nature's.

All this is true in motivational terms, though in teleological terms the technology of naturoids could be understood as a matter of a form of control, but of a higher and more ambitious kind since one expects to substitute the highest possible form of command, that is the capability to create nature *ex novo*.

Nevertheless, conventional technology, with its materials and techniques, poses serious limitations on the technology of naturoids. In fact, on one hand the technology of naturoids allows us, at certain observation levels, to 'deceive' an organism, a spectator, an environment, a living species or some structures or natural events. On the other hand, it re-proposes, at different levels and often also at the one assumed by the designer, its own heterogeneity compared to the exemplar and its performances. Hence, in spite of the designer's aims, naturoids tend to create realities which deviate more and more from nature, just as conventional technology does deliberately right from the beginning.

M. Negrotti, *The Reality of the Artificial*, Studies in Applied Philosophy, Epistemology and Rational Ethics 4, DOI: 10.1007/978-3-642-29679-6_23,

© Springer-Verlag Berlin Heidelberg 2012

This does not mean that naturoids are intrinsically destined to fail: in fact, we know that they have had some success. However, in producing a naturoid, one has to keep in mind that the similarity with what happens in the natural exemplar will always be accompanied by many unavoidable effects or properties whose interplay with the selected essential performance may generate a 'new version' of the natural phenomenon.

Anyway, we can maintain that the technology of artificial reproduction of natural instances has its own unique philosophy which differentiates it from conventional technology, above all in terms of ideation, design and constraints. In particular, we have established the following points:

A naturoid always derives from a process of multiple choices which, as such, prevents it from pursuing the overall analytical reproduction of the natural object or process which it intends to reproduce. These choices pertain to the observation level, exemplar and essential performance.

The selection of an observation level, which in some fields, such as art, includes the possibility of deliberate construction without any direct reference to the sensible world, leads to the formation of individual or collective representations or models—of the natural object observed—which do not necessarily respect 'reality as it is'. Representations may depend on a whole series of cultural premises, values, preferences, beliefs or prohibitions, of which history is full.

The selection of the exemplar, in turn, inevitably assumes the form of an isolation, and at times a real eradication, of the object or process from the natural context in which it is found. Fixing the boundaries of the exemplar is always an arbitrary operation whose success, as a rule, cannot be easily predicted.

In some cases, the resulting model does not have serious consequences for the reproducibility of the exemplar at the chosen observation level. In many other cases, however, its isolation cuts off, so to speak, relationships at various observation levels with the rest of the context to which it belongs. These relationships could be and often are vitally important in determining the characteristics or the behavior of the natural object in its natural context, and, therefore, their reduction within the model may seriously affect the characteristics or the behavior of the resulting naturoid.

The selection of the essential performance is already, on the one hand, bound by the selection of the observation level and exemplar and, on the other hand, almost always constitutes a kind of 'bet' on the quality, function or behavior which is considered 'fundamental' in the exemplar. Furthermore, the selection of the essential performance implies another process of isolation of that performance from the other performances or properties which characterize the exemplar's way of being in nature. In other words, the selected essential performance is, almost always, unavoidably modelized as a 'purified' performance, i.e., as if it were a 'context-free' property or behavior.

The principle of 'functional equivalence'—between the exemplar and its formal model or its concrete reproduction—only guarantees that, in a few circumstances, the essential performance is sufficiently autonomous and therefore transferable to different supports. But this fact does not guarantee that such supports reproduce the structure of an exemplar and, even less, that the neglected performances of the exemplar will spontaneously 'emerge'. Hence, there is no reason for the complete set of properties of a naturoid—apart from the selected and reproduced essential performance—to overlap that of the natural exemplar.

The emergence of properties, qualities and behavior due to the "nature of naturoids" and not due to the nature of the exemplars, is thus inevitable for the technology of naturoids. These "new" properties, which are not explicitly planned in the design, can include realistic but unplanned performances of the exemplar if, and only if, these are close functions of the planned essential performance. For instance, if one designs a doll that will cry when a child strikes it, then it is also very likely that the doll will cry even if it also falls on the floor, just like a real baby.

Compatibility with the host context, environment or organism is usually one of the most crucial accomplishments for naturoids, since it is precisely at that stage that the heterogeneity of its structure or its performances will clearly emerge. In fact, in order to have a naturoid performing *exactly* like the exemplar, we should be able to design the whole natural context within which it lives in nature.

Deceit, simulation and illusion can certainly be usefully pursued using various strategies, but the inevitably rigid limits within which they are possible rely heavily on the essential performance's autonomy from the 'support' which generates it in nature. At times this autonomy can be relatively high as it is when we are dealing with exemplars or, even more so, with purely informational performances, while it is usually very low in all the other concrete cases.

The synthesis or coordination of more than one partial naturoid (each based on its own exemplar and essential performance) creates further problems. Indeed, putting together two artificial objects or processes only increases the distance of the presumable resulting object or process from the system made up of the two exemplars in nature. This is mainly due to the relationships, within the natural exemplar, among structures and processes at observation levels which are inevitably neglected in the model which guides the reproduction. This could be called the 'unavoidable price of analysis'.

The design of a *third level coordinator* of two naturoids can be conceived as an expedient or as a quite new project of a naturoid. While in the former case the resulting object or process will have no intended similarity with what is found in nature, in the latter case we are dealing with quite a new undertaking. Indeed, the reproduction of the coordinator module could require giving up or changing the observation levels assumed for designing the first two naturoids. In turn, this could involve changes in the structure or the performances of the two partial naturoids involved, so that they fit the requirements of the third observation level. Thus, a sort of *ad infinitum* rebound between bottom-up and top-down strategies could take place.

In any case, though today the synthesis of two or more naturoids seems impossible if one expects the same performances we get from the natural exemplar (above all the performances that were not explicitly designed), we should not overlook the fact that *something real always happens* in such attempts.

Therefore, the transfiguration of the exemplar and its performances, and often the essential performance as well, constitutes an inexorable tendency of naturoids. It is due to the combined effect of the selective and heterogeneity factors mentioned above. Since nature, including human beings, is a whole which is intrinsically integrated—by physical, chemical, biological, psychological, and sociological laws—even the optimal reproduction of one performance of an exemplar already constitutes an anomaly in itself, because it does not have, nor can it easily have, natural relationships with the host context at all the natural levels involved when we consider it in nature. Anyway, transfiguration does not always imply a failure of naturoids, because often we derive from the design process some new ideas or useful things.

No matter what actually happens and will happen through the research which is planned in many fields of naturoids, we can be certain that from whatever perspective they are examined, they tend to generate a separate reality, which should be studied in depth, and will be in the coming years. Indeed, the density of naturoids, processes and machines is increasing steadily, thanks to more refined technologies which allow man to renew his utopian visions, and because many aspects of our existence have always been ready and available for gaining practical and imaginative, scientific and artistic benefits from those technologies.

It is likely that the dream of realizing a cyborg (cybernetic organism, that is, a natural man who is improved and amplified in his performances by artificial devices connected directly to his body) will remain unattainable for a long time, if we take the exaggerated characters of science fiction films as examples. Nevertheless, as we have seen, today millions of people live, work, communicate or move thanks to extensions, prostheses or implants of naturoids of various kinds. The prospect of a more marked physical integration with the artificialized world is therefore not unthinkable. However, serious sociological, psychological and even ethical problems are already appearing on the scene. We do not know exactly what this will entail. Whoever claims to know what the future in the field of naturoids will be really should not be taken seriously.

The only issue that we should not forget is that our species is not willing to accept all that is technologically possible, mainly if the transfigurations of the performances of our body, mind and environment will bear innovations that appear unmanageable and conflict with our preferences. Anthropological culture, in this sense, acts as an extraordinary filter that accepts or refuses according to criteria that cannot be indefinitely transformed.

Appendix A: Naturoids and Music[1]

Reproduction and Transfiguration

According to a well-known anecdote, the music critic Pierre Lalo, after having listened to *La Mer* by Claude Debussy, declared "I have the impression of beholding not nature, but a reproduction of nature, marvellously subtle, ingenious and skilful, no doubt, but a reproduction for all that … I neither hear, nor see, nor feel the sea" (Vallas 1973). As François Lesure noted, Debussy was trying to express, not the image of the sea, but, rather, its recomposed memory. After all, it is known that Debussy saw music as an art which, perhaps more than any other, is not intended to reproduce nature exactly, but which instead aims at the 'mysterious affinity between nature and imagination.' This brief remark could prelude, as has often happened, some subtle analyses of the symbolic or descriptive character of music.

For my part, I would like, instead, to place the matter of the reproductive capacity of music in the more general context of mankind's attempts to reproduce objects observed in nature, that is to say, to realize naturoids.

It almost goes without saying that I shall assume that musical composition, in common with all artistic composition, possesses all the characteristics of an attempt to reproduce something that, in whatever way we choose to define it, is kept—or, rather, is generated—in the composer's mind. After all, if art always includes an expressive component, then the artist externalizes or *expresses* something; that is to say, recalling the Latin roots of the term, he presses something out from himself.

Debussy's above-cited affirmation is, in this regard, undoubtedly central, because it is the *image* of a natural phenomenon, and surely not the phenomenon *in*

[1] An early version of the content of this appendix was presented at the International Conference on the Emotional Power of Music, Geneva, June 2009.

M. Negrotti, *The Reality of the Artificial*, Studies in Applied Philosophy, Epistemology and Rational Ethics 4, DOI: 10.1007/978-3-642-29679-6,

© Springer-Verlag Berlin Heidelberg 2012

itself, which constitutes the starting point of any reproductive enterprise. I do not know if the expression 'absolute music'—which refuses the descriptive character a composition may have—is a founded one, but, surely, when the composer's image is transduced into music it becomes really another thing, even when he aims to reproduce a natural phenomenon. Rightly it has been reported that Beethoven had "always a picture in mind" when he was composing (Knight 2006), but, when he wrote remarks on the score upon his Pastoral symphony, he reminded the oboe section that music is not painting. On the other hand, as D.B. Knight recalls, even R. Schumann denied having natural phenomena in mind when composing, though he gave some of his compositions titles referring to nature. All this lets open the door to the hypothesis according to which the pictures composers have in mind, though they could be related to some sense experience, are of a totally new stuff, namely strictly musical (Knight, p. 42).

In the field of technology, from a methodological point of view, anyone who wishes to reproduce a natural object or process—be it, say, a flower, an arm, intelligence, or the sense of smell—cannot but establish a *model* of the entity to be reproduced. After all, the concept of *model* absorbs and generalizes that of *image*.

Even in operational terms, the concept of model or of image clearly establishes that any reproduction is not a mere two-phase process—not simply the observation of the world followed straightforwardly by its reproduction. On the contrary, there is a third, decisive, intermediate phase, which is that of modelization.

It is necessary to remember that a model is, in turn, the result of a complicated process which, beginning with the interaction with an object or phenomenon, leads to its description according to some observation level. As a logical consequence, a model will allow—and, indeed, will impose—the description of the observed object only through categories and properties that are compatible and consistent with the particularity of the adopted observation level. For instance, the description of a tree from a mechanical observation level will bring to the foreground features that differ greatly from those brought forward when adopting, say, a biochemical observation level.

A work of art is always something 'other' than the natural object, not only because of the recourse to materials that are different from those that nature adopts, but also, and perhaps even more so, precisely because nature is already literally transduced into the image, thereby becoming a different reality which appears strongly reduced and polarized when compared to the natural exemplar, owing to the adoption of a certain observation level. This is true even when the composer does not observe, or cannot observe, directly what he wants to reproduce and draws the exemplar from an external account, as in the case of *The Creation* by F.J. Haydn that follows the description of the Creation in the Bible. The same has occurred in many other cases and also in the history of painting, of course. Anyway, what we hear and what we see are not storms or floods but the image of them in the composer's mind.

Nevertheless, in the third phase of the reproduction process, which includes the actual reproduction, the recourse to concrete factors, such as matter and energy, will bring the work back into the real world, assigning to it its own high-level materiality, whose complexity may, at times, match, or perhaps even exceed, that of the exemplar on which it is based.

The illusory character of a work of art, as underlined by Susanne Langer (1953), for example, is only a sensible correlate of art conceived as a reproductive enterprise. It is a sort of residual extension, inherent in the work of art, of the sense experience of the artist, which, after being processed within the image, remains, in greater or lesser measure, in the final product, without having any aesthetical relevance in itself.

We might say that the extra-musical content is intended to reproduce, not an *illusion*, but rather an *allusion* to existential or natural factors that become immediately surpassed by, and recomposed in, the artistic composition. This is why, for me, Ottorino Respighi's introduction of recorded birdsong into his *Pini di Roma* is particularly instructive, in that the definite separation between the natural phenomenon, in its original form, and the accompanying music, clearly indicates the distance between the sense experience and the artistic reproduction.

In this regard, an interesting analogy holds good for the technological reproduction of natural exemplars. In fact, even technological naturoids, as we have seen, aim to "cheat the body", because whatever the context within which the naturoid is placed, it has to be persuaded to have to deal with a natural device. In fact, the parts of the body that receive the blood pumped by an artificial heart are interested, so to speak, only in a haematological observation level and nothing else.

Of course, the fruition of a work of art—be it pictorial, musical or whatever—is not aimed at cheating an organism in order to get some special response from it, even if the history of art is not lacking in examples of descriptivism which seek to involve people in various realistic ways. Beyond figurative painting, with its various religious or secular objectives, we need think only of theater scenography, or the numerous *trompe l'œils*, or, in music, not only the descriptive tradition, but also the deliberate onomatopoeics. Among the most important examples we find *The Four Seasons* by A. Vivaldi, the twelve pieces *The Seasons*, for piano, by P.I. Tchaikovsky, *The Moldavian* by B. Smetana, *Musical Portrait of Nature* by J.H. Knecht, the concertos *The Hunt, The Night* and *The Goldfinch* by Vivaldi, *The Sea-Storm* by I.J. Holzbauer, the famous pieces for harpsichord *Les papillons, Le moucheron, Le rossignol en amour* by F. Couperin, and the pieces by P. Rameau *Le rappel des oiseaux, La poule*. Other noteworthy examples include Symphony No. 6, The Pastoral by L. van Beethoven, *The Fountains of Rome* and *The Pine Trees of Rome* by O. Respighi, who also wrote *The Birds*, a suite for small orchestra, *Saudades das Selvas Brasileiras* and *Alvorada na Foresta Tropical* by H. Villa-Lobos and the compositions *Réveil des oiseaux* and *Oiseaux exotiques* by O. Messiaen. The skillful presentation of naturalistic effects at the beginning of *Peter and the Wolf* by S. Prokofiev is also famous. Even Johann Sebastian Bach gives us a masterpiece of this kind with the Capriccio in B-flat, "On the departure of a beloved brother" (BWV 992).

The point is that, in the technological reproduction of a natural object, realism is compulsory, even if often very difficult to achieve. In such circumstances, the illusion, the deception, embedded in a naturoid is intentionally pursued in order to allow some performance in a given natural context, which is always ready to reject that which is not homogeneous to it.

In bioengineering, the deceit is omnipresent almost by definition, and is welcome, of course. In its more advanced projects, as in the field of artificial organs, for instance, it clearly exhibits this tendency.

We may also refer to the huge quantity of naturoids designed to fill lacks in various naturalistic situations, such as artificial nests or artificial reefs, destined to support biological species whose survival would otherwise be endangered. In such cases, the illusion, or the deceit, is known to the designer, but goes almost unnoticed by the natural system involved, as long as such a system assumes the same observation level as that of the designer. I say 'almost unknown' because, unfortunately, nature is not so easy to cheat, and, sooner or later, in many cases, it will discover the extraneousness of the artificial, and present the bill, so to speak, usually in the form of rejection or unforeseen side effects.

As a result of these unplanned outcomes, the technology of naturoids so often proves unsatisfying, and perceives the inefficacy of the deceit as a failure of the models themselves.

Regarding these themes, I have defined as 'transfiguration' of natural exemplars the inevitable difference that any artificial device exhibits with respect to the natural instance. One might say that technological research, though not always, pursues the minimization of such transfiguration, and intends to proceed in exactly this direction, even if the goal always moves ahead as it is approached.

By contrast, art follows a different direction in an opposite direction, as it has no actual benchmark to exceed within natural reality. The image generated by the artist is always, so to speak, precise and complete, because the artist has nobody to deceive, although the deceit is technically possible, and, in some historical eras, has been intentionally pursued in the form of the above-mentioned onomatopoeic experiences.

In other words, the fidelity of the image with respect to the natural exemplar is not the artist's true objective. Rather, he or she seeks to reprocess the sense data according to his own poetics; that is to say, by producing a deliberate transfiguration of the exemplar. In this regard, Gustav Mahler provides, in my opinion, a clear and almost conclusive pearl of wisdom when he states that "If a composer could say what he had to say in words, he would not bother trying to say it in music."

In fact, in order to communicate an emotion—i.e., in order to make it common to my mind and someone else's—adopting a musical language would amount to taking the most difficult, winding and often misleading road. Not by chance does the understanding of the possible extra-musical content of a non-onomatopoeic composition, particularly when it has no explicit title, and even if accompanied by words, almost always requires very complex analysis. Such would be the case for Franz Schubert's *Lieder*, for example, or for the reconstruction of pictorial affinities in *Paintings at an Exhibition* by Modest Mussorgsky.

We may set up an analysis which exploits a sort of 'reverse engineering' aimed at revealing the natural phenomena that may be embedded in the music. But this is an operation that leads ineluctably to the separation of the two levels, the musical and the natural, in the tacit—and, in my view, unfounded—conviction that the beauty of music becomes amplified by the recognition of its possible naturalistic substrate.

Perhaps flying in the face of common belief, I maintain that Igor Stravinsky was right when he declared "For its own nature, music cannot explain anything: neither emotions nor points of view; neither feelings nor natural phenomena. It can explain only itself."

The Rarefaction of Meaning

We may summarize the above discussion as follows. If one assumes the centrality of the concept of reproduction of mental models or images in communication, art and music, then one must confront the issue of observation levels, because any model is strongly affected by the choice of such a level. This conclusion is taken as a matter of fact in every area of the technology of naturoids.

The objective of any reproduction process, be it in the field of technology or of communication, consists in the achievement of the highest possible level of realism. This means that the reproduced object or process should be as faithful a copy as possible of the exemplar.

This objective is pursued not only through suitable technologies or languages, but also, and above all, by reducing the holistic reality of the natural object to the form of a phenomenology polarized around a dominant profile or level of observation. Nevertheless, the resulting reproduction never amounts to an actual replication, as this is prevented not only by the adoption of different materials and procedures than the natural ones, but also by the selection of one only observation level at a time, from among the infinity of possible levels characterizing the ontology of any natural phenomenon.

The transfiguration—that is to say, the discrepancy between the properties of the natural object and those of the artificial one—assumes, in the world of machines, the form of malfunctions or side effects, while in the world of communication phenomena it takes the form of equivocations, improper mental associations, and sometimes almost complete lack of understanding. This breakdown of effective communication is rendered all the more probable when a large number of interlocutors do not share compatible observation levels regarding the topic under discussion.

However, if we turn our attention from ordinary communication to art, we may ascertain that the reproductive objective of the artist, which actually exists, and which leads to expression, does not concern the sense content of his human experience, but, rather, the image of the experience generated within his or her mind. In other words, in the technological field the model constitutes a tool constantly in need of empirical, and, so to speak, external verification, whereas in art the image constitutes the originating exemplar, i.e., the reality that is to be reproduced. In some measure transfiguration happens also in our everyday communication since, whenever we build a message by means of the "linguistic technology" available, we stop the building process when the message *appears to us* as able to reproduce our mental state in a sufficient measure. Really, we are the first listeners of ourselves and this is why our inability to sometimes express ourselves leads us to make statements like "I do not know how to tell you", "I have no words" or "more clearly than that." This is a situation that the Italian dramatist L. Pirandello sums up well when he declares that

> in the passage from one mind to another, modifications are inevitable ... it rarely happens that a writer is pleased that his work, for a critic or for a reader, remains that same work, or thereabouts, that he expressed, and not another, ill-considered and arbitrarily reproduced (Pirandello 1908)

More markedly, the artist also will stop his work when it will seem to him as close as possible to his own image of the world and not when he feels that everyone can recognize what is represented, as technologists or communicating people do.

A technological naturoid comes from the objective observation of a natural exemplar, and aims at its reproduction in equally objective terms, in such a way that it might be recognized, in its essential performance, as a partial or complete reproduction of the original. To achieve this, designers must limit the transfiguration effect as far as possible. Thus, for example, an artificial heart must exhibit at least some of the characteristics and essential performances of a natural heart.

By contrast, a work of art, even when it starts from a naturalistic objective observation, not only does not fear the transfiguration, but even tends to assume it as the main objective, in accordance with the aesthetic vision of the artist. The ambiguity of every work of art, or the opening towards alternate interpretations, is also based on this objective. To put it differently, while technological reproduction and ordinary communication tend towards convergence of result and sensible reality, a work of art generates divergence; and this explains why, as Massimo Mila has written,

> We accept the Debussy of Gieseking and that of Cortot, the Chopin of Paderewski and that of Rubinstein, the Beethoven of Fürtwängler and that of Toscanini (Mila 2001)

Therefore, the transfiguration of an empirical, sensible reality occurs just at the moment of the formation of the images of the world, exactly as if the artist, in experiencing the world, assumed an observation level already dominated by his aesthetical orientation. This seems to be consistent with, among other things, Levi-Strauss's remark concerning the opposition between myth and music, according to which the former is free from the sound but is linked to the sense, while the latter is free from the sense but is linked to the sound. Similarly, Boris de Schloezer maintains that music, thanks to the overlapping of signifier and signified, has no sense because it *is* a sense (Fubini 1995).

The ambiguity and the multiplicity present in the work of art are not the sign of a failed reproduction but, rather, the result of the image of the sense reality arising in the artist's mind after having adopted a *sui generis* observation level. It is a matter of an observation level that acts as if it were a new, culturally established sense organ added to the natural set, and this explains why the appreciation of a musical piece, not by chance, needs several, accurate listenings and induces different interpretations.

This musical observation level, as we might call it, is so powerful that it can allow the composition or appreciation of music even in the absence of sounds, or, as in the case of Ludwig van Beethoven, when one has a progressively weakening sense of hearing.

Furthermore, the allusion to naturalistic elements can even produce moments of collision or interference with the strictly musical objectives of the composer, as happened in the case of the 'damned bird' which interfered with Antonin Dvorak's work at Spillville. Dvorak reproduced the bird's singing, but by means of expressive modalities that were much more than pure imitation, of course.

The transfiguration due to the observation level appears all the more clearly in the case of non-onomatopoeic music, however, because the recognition, by a typical member of the public, of the natural exemplar (be it the sunset, the waves of the sea, a landscape or an emotion) is almost always impossible, even if he is able to grasp the expressive power of the composition, thus ensuring the flow of at least something from the author to the listener.

In painting, where the representational tradition has been much more intense and lasting, we can speak, likewise, of a pictorial observation level, but the so-called 'rendering' of a work almost always consists in the transfiguration of at least a sufficiently recognizable core. Music, in turn—perhaps because it is formulated in a language that has little to do with the world of meanings, and that is centred more on the form, and therefore on syntax, rather than on 'translating' the author's emotions—constitutes its own observation levels from the outset, and presents them in terms of an irreducibly musical emotionality. Adopting a rather forced analogy, we might say that, if music *were* painting, it would surely be, in the vast majority of cases, abstract painting; or at least it would appear as such if the listener claimed to recognize the reproduction of elements of shared sense experience. As has been correctly said,

> … musical descriptions function much more like descriptive *nonrepresentational* paintings than like representational paintings. In short, music can *describe* but only rarely *represents* the world (Robinson 1987)

We need add only that the fruition itself of music generates emotions, of course, but that the deepest emotions intervene when the listener also succeeds in placing himself at a strictly musical observation level, neglecting the extra-musical effects linked to certain aspects of his own personality, or to situations or memories associated with past listening.

Regarding this point, there is a strong controversy between the cognitivist and emotivist schools of thought, and I cannot but agree with Jenefer Robinson when she observes that

> the emotions aroused in me are not the emotions expressed by the music (Robinson 1994)

However, the difference is only a presumed one, as it cannot directly be verified at all, and any position in this regard must count on the strength of the theoretical argumentation, or on some kind of mental experiment.

Let us take an imaginary situation, which we could name the 'crying Korean.' I do not know the Korean language, but let us imagine that I meet a young Korean who is crying, holding his face with his hands, uttering phrases with an evidently sad and woeful pitch. Although I cannot understand the nature of his pain, surely I shall experience feelings of compassion. Yet my feelings and his will be completely different. In the same way, the emotional effects of music develop even if the composer and the listener do not share the same observation level. For this reason too, therefore, the effects of music should not be confused with its most genuine appreciation.

Anyway, it should be remembered that the emergence of peculiar observation levels, or of 'senses' which add themselves to the physiological ones, is not an isolated fact, present only in the arts. The mathematical or geometrical observation of the world, for example, constitutes a dramatic human construction, which, from Pythagoras to Mandelbrot, has made available a profile of reality that goes beyond our usual way of looking at things. The same might be said for the systemic or informational visions of the empirical world. They are true cultural inventions that do not overlap the special 'optics' of the theoretical–experimental sciences. These sciences, indeed, along with the observation levels that they imply (e.g., the physical or the biological, the psychological or the ecological), deal with empirical reality assumed as datum, while the intentionally built levels I have referred to, including the musical one, transcribe reality by means of their own, literally invented, grammar and syntax, offering themselves as additional instruments for describing the world, or for its poetical transfiguration.

However, it should be admitted that, in the analysis proposed here, the aptitude of music to construct, in the composer's mind, its own observation level is the result of a logical deduction on my part, and not yet of an experimental verification. The sequence of my assertions may be synthesized as follows. It forms a sort of pyramid as a metaphor of the incremental loss of ordinary sense and of the parallel increment of specific sense in the transition from communication to art, and from art to music.

Any form of communication consists of a reproduction of mental states, and therefore generates artificial states—i.e., so-called messages. Successful communication is possible only when the receiver of a message places himself at the same observation level as its sender.

Art is a kind of communication, and thus it also generates artificial objects, although they are, to a greater or lesser extent, intentionally transfigured on the basis of the image the artists builds in himself on his own observation level. On the other hand, at least in the case of figurative art, anyone who enjoys a work of art is sufficiently able to share the observation level of the author exactly because the artificially reproduced sense references usually remain recognizable.

Music, in its turn, is surely a kind of art, and therefore, too, a kind of communication, but the observation level of the author is not easily isolatable and sharable. Indeed, in the final 'naturoid' generated by the music, the references to the sense experience, even if often indicated explicitly in the titles, widely escape our perception.

Nevertheless, there is no doubt that the composer lives in the sense reality, drawing from it mental states and emotions that he then seeks to reproduce musically. The fact that we cannot easily recognize the sense references in the final artwork means, therefore, that the composer resorts to a non-ordinary observation level that may be defined as *intrinsically musical*.

At this point, it could be interesting to consider the following quotations, assuming them as quasi-empirical support concerning the nature of musical composition. For the most part, they are very well known, but, in the context of the present discussion, they may perhaps exhibit a special flavor, because, beyond their rhetorical character, they reveal some of the elements (signalled in boldface type) that I have introduced here in a purely theoretical way. The translations, where they have proved necessary, and in boldface, are mine.

Statements by Writers or Scholars in the Humanities

It is worth noting the insistence on the concepts of 'invisibility' and of 'alterity' assigned to what music expresses. Nevertheless, according to the perspective developed above, music produces what is perfectly visible to the composer at his own observation level. The alterity itself of the musically described world is such only for the observation modalities that music makes possible. It is a level that is accessible, at least partially, to whoever possesses not only a necessary sensitivity to this kind of art, but also sufficient notational, instrumental and orchestral competence.

In the end, other philosophers and scientists, such as Henri Bergson and Albert Einstein, were right when they said, respectively, that "The eye sees only what the mind is prepared to comprehend" and "We see the world through our theories."

Leonardo da Vinci: "The poet ranks far below the painter in the representation of **visible things**, and far below the musician in that of **invisible things**."

D. Diderot: "Painting is a more natural art, whereas music is an art very much more **linked to man**."

V. Hugo: "Music expresses **that which cannot be said** and on which it is impossible to be silent."

L. Tolstoy: "Music is the shorthand of emotion."

H.C. Andersen: "Where words fail, **music speaks**."

V. de Laprade: "It is undeniable that music induces in us a sense of the infinite and the **contemplation of the invisible**."

E. Sapir: "[Music represents] a **more difficult mental level**, more **elusive** than expression itself."

J. Blacking: "The Balinese people speak of '**the other mind**' as a state of being that can be reached through dancing and music."

A. Huxley "After silence, that which comes nearest to **expressing the inexpressible** is music."

Statements by Composers

The theme of alterity arises among composers also. The visibility, on the contrary, assumes the contour of a perception reserved to the composer, who is able, in other words, to observe the world musically.

L. van Beethoven: "Music is the one **incorporeal entrance** into the higher world of knowledge which comprehends mankind but which **mankind cannot comprehend**."

F. Mendelssohn: "Though everything else may appear shallow and repulsive, even the smallest task in music is so absorbing, and carries us so **far away from** town, country, earth, and **all worldly things**, that it is truly a blessed gift of God."

J. Brahms: "[N]ot only do I **see distinct themes** in my **mind's eye**, but they are clothed in the right forms, harmonies, and orchestration."

I. Stravinsky: "By itself, music cannot explain anything: neither emotions, nor points of view, nor sentiments, nor natural phenomena. It can explain nothing but itself."

F. Busoni, on Mozart: "His palace is immeasurably great, but he never steps outside the walls. Through the windows of it **he sees nature**. The window frame is also the frame of nature."

H. Walcha: "Bach **opens a vista** to the universe. After experiencing him, people feel there is meaning to life after all."

On the basis of the above quotations, one could build the following synthetic definition of music. It might seem a sort of joke, but, in the context of our discussion, it assumes a special conceptual flavor.

> Music is an art very much linked to man and it deals with the other mind, to which it gives the possibility to get an incorporeal entrance in the contemplation of the invisible, far away from all worldly things, for expressing the inexpressible. Nevertheless, the composer opens a vista, he sees nature and sees distinct themes in his mind's eye. But the music cannot explain anything: neither emotions, nor points of view, nor sentiments, nor natural phenomena: music speaks at a more elusive and difficult mental level, of what cannot be said by words and mankind cannot comprehend.

Appendix B: Naturoids and Conventional Technology Devices

Generally, the many kinds of artificial devices normally defined as automatisms are excluded. Moreover, some of the objects or processes which are defined here as artificial do not correspond with the definition of naturoid introduced in this book. On the contrary, they constitute the result of a patchwork or recombination of the same materials present in their exemplars.

Some of the devices are reported without the name of their inventors because of their very ancient and diffuse existence or because of their difficult, precise attribution.

Before 1800 (Various Sources)

Artificial cornea (G. Pellier de Quengsy)
Artificial fireworks
Artificial flower
Artificial horizon (J. Elton)
Artificial ice
Artificial insemination (L. Spallanzani)
Artificial irrigation
Artificial island
Artificial lake
Artificial marble
Artificial propagation (agricultural botany)
Artificial rainbow (F. Bacon)
Artificial selection

M. Negrotti, *The Reality of the Artificial*, Studies in Applied Philosophy, Epistemology and Rational Ethics 4, DOI: 10.1007/978-3-642-29679-6,
© Springer-Verlag Berlin Heidelberg 2012

Artificial writing (J. Gutenberg)
memoria artificiosa (Cosmas Rossellius)
perspectiva artificialis (L. B. Alberti, P. della Francesca)

After 1800 (Various Sources)

Artificial adaptation (J.H. Holland)
Artificial arm (V. Kolff)
Artificial bait
Artificial barrier
Artificial beach
Artificial bells
Artificial blood (R. Naito)
Artificial blood vessel (A. Carrel)
Artificial blue (J. Guimet)
Artificial bone
Artificial brain cell (A. Richter-Dahlfors)
Artificial cardiac pacemaker (J. Hopps)
Artificial cardiac valve (C.A. Hufnagel)
Artificial cavity
Artificial cell (T.M.S. Chang)
Artificial colours (A. Baeyer)
Artificial diamond (P. Williams Bridgman)
Artificial drying
Artificial ear (A. Djourno, C. Eyriès)
Artificial environment
Artificial esophagus (H. Neuhof)
Artificial experts (M.H. Collins)
Artificial extremities
Artificial eye
Artificial fertilizers (J.B. Lawes)
Artificial fibers
Artificial fish (M.S. Triantafyllou)
Artificial flavour
Artificial gelatin
Artificial grass (D. Chaney)
Artificial gravity
Artificial ground
Artificial habitat
Artificial hair
Artificial hand

Artificial hatching
Artificial heart (P. Winchell, W. Kolff, D. Liotta)
Artificial hip (T. Gluck)
Artificial honey
Artificial horizon for airplanes (A. Sperry)
Artificial incubation
Artificial intelligence (M. Minsky, H. Simon et al.)
Artificial island
Artificial ivory (H. Scarton, S. Calabrese)
Artificial joints
Artificial kidney (W. Kolff)
Artificial knee (A. Burstein)
Artificial lake
Artificial landscape
Artificial language
Artificial larynx (J.E. Mackenty)
Artificial leather
Artificial leaf (D. Nocera)
Artificial leaves (S. Winkler)
Artificial life (C.G. Langton)
Artificial ligament
Artificial light (T. Edison)
Artificial limbs
Artificial liver (K. Matsumura, A. Demetriou)
Artificial lung (G. Mortensen)
Artificial milk (J. von Liebig)
Artificial muscle
Artificial musk (A. Baur)
Artificial nail (F. Slack)
Artificial nest
Artificial nerves
Artificial organs
Artificial pancreas
Artificial paradises (C. Baudelaire)
Artificial pearl (M. Koukichi)
Artificial perfume
Artificial plant
Artificial pond
Artificial radioactivity (F. Joliot)
Artificial rain (I. Langmuir, V. Schaefer)
Artificial reality (M. Krueger)
Artificial rearing
Artificial reef
Artificial resin
Artificial respiration

Artificial retina (A. Chow, V. Chow, M.A. Mahowald, C. Mead)
Artificial rock
Artificial rubber (W. Carothers, J. Nieuwland)
Artificial satellite
Artificial sensors
Artificial silk (H. de Chardonnet)
Artificial skeleton
Artificial skin (J. Burke, I. Yannas)
Artificial smell (Sony Corp.)
Artificial snow (Emile Wyss & Cie SA)
Artificial sound
Artificial speech
Artificial star
Artificial starch
Artificial sweetener
Artificial tanning
Artificial taste
Artificial teeth
Artificial tears
Artificial weathering
Artificial wood

Examples of Conventional Technology (Starting from the Sixteenth Century)

Medical thermometer (Santorio, Galileo)
Telescope (Lippershey)
Printed newspaper (Avisa—Relation oder Zeitung)
Barometer (Torricelli)
Pendulum clock (Huygens)
Pressure cooker (Papin)
Steam engine (Savery)
Flying shuttle (Kay)
Electrical capacitor (von Kleist and van Musschenbroek)
Motor vehicle (Cugnot)
Streetcar (Outram)
Centrifugal governor (Watt)
Electrical cell (Volta)
Carbon paper (Wedgwood)
Bicycle (von Drais)
Stethoscope (Laënnec)

Steering wheel (Ackermann)
Piano (Demian)
Electrical engine (Henry)
Relay (Henry)
Revolver (Colt)
Ship propeller (Ericsson)
Fax (Bain)
Vulcanization of rubber (Goodyear)
Telegraph (Morse)
Nitroglycerin (Sobrero)
Saxophone (Sax)
Gyroscope (Foucault)
Elevator (Otis)
Storage battery (Plant)
Linoleum (Walton)
Rotary press (Hoe)
Machine gun (Gatling)
Generator (Pacinotti)
Dynamite (Nobel)
Celluloid (Hyatt)
Compressed air brakes (Westinghouse)
Chewing gum (Semple)
Drill (Morrison)
Telephone (Meucci)
Gramophone recorder (Cros)
Electric lamp (Edison)
Cash register (Ritty)
Cable railway (Olivieri)
Power station (Edison)
Fountain pen (Waterman)
Steam turbine (Parsons)
Linotype (Merenthaler)
Aluminum (Hall and Hiroult)
Straw (Stone)
Tire (Dunlop)
Eiffel Tower (Eiffel)
Electric chair (Brown and Kennelly)
Electromagnetic wave detector (Branly)
Thermos (Dewar)
Reinforced concrete (Hennebique)
Escalator (Reno)
Zipper (Judson)
Sphygmomanometer (Riva-Rocci)
Slot machine (Fey)
Magnetic recorder (Poulsen)

Disk brakes (Lanchester)
Diode (Fleming)
Joystick (Esnault-Pelterie)
Vacuum cleaner (Spengler)
Cellophane (Edwin)
Neon tube (Claude)
Toxic fume suppresser (Frenkel)
Radioactivity counter (Geiger)
Brassière (Jacob)
Pyrex pot (Littleton)
Tank (Swinton)
Mass spectrograph (Aston)
Cement (Dickson)
Electroencephalograph (Berger)
Television (Baird)
Combustible fluid for missiles (Goddard)
Radio compass (Busignies)
Helicopter (D'Ascanio)
Nylon (Carothers)
Lie detector (Keeler)
Instant coffee (Nestlé)
Spray (Kahn)
Jet (von Ohain)
Atomic battery (Fermi)
Ball-point pen (Bíró)
Atomic bomb (Oppenheimer)
Microwave oven (Spencer)
Radial tires (Michelin)
Pinball machine (Mabs)
Transistor (Shockley, Brattain, Bardeen)
Tetra-pack (Rausing)
Solar battery (Pearson)
Optical fiber (Kapany)
Integrated circuit (Kilby)
Hovercraft (Cockerell)
Laser (Maiman)
Tape cassette (Philips)
Crystal liquid monitor (Heilmeier)
Microprocessor (Faggin, Hoff, Mazor)
CAT (Hounsfield)
Modem (Hayes)
Floppy disk (Apple, Tandy)
Compact disk (Philips, Sony)

References

Aben, R., de Wit, S.: The Enclosed Garden History and Development of the Hortus Conclusus and its Reintroduction into the Present-Day, Urban Landscape, p. 202. 010 Publishers, Rotterdam (1999)
AFCEA International Press, p. 60 (1988)
Arnheim, R.: Entropy and art: an essay on disorder and order (Entropia e arte, Einaudi), p. 53. University of California, Torino (1971)
Aunger, R.: What's special about human technology. Camb J Econ **1**, 7 (2009)
Baggi, D. (ed.): Readings in Computer Generated Music. IEEE, Los Alamitos (1992)
Balasubramanian, D.: New eyes for old? The Hindu, Adapted from the Convocation Address given at the Elite School of Optometry, Chennai, 5 Sept 1998
Bass, M. (ed.): Handbook of Optics III: Vision and Vision Optics, p. 21. McGraw-Hill Professional, New York (2010)
Basti, G.: Intentionality and foundations of logic: a new approach to neurocomputation. In: Kitamura, T. (eds.) What Should Be Computed to Understand and Model Brain Function?, p. 251. World Scientific, Singapore (2001)
Bateson, G.: Steps to an Ecology of Mind, p. 250. Chandler, San Francisco (1972)
Bedini, S.A.: The role of automata in the history of technology. Technol Culture **5**, 1 (1964)
Benyus, J. http://en.wikipedia.org/wiki/Biomimicry (cite_ref-Benyus_1997_6)
Benyus, J.: Biomimicry Innovation Inspired by Nature. William Morrow & Company, Inc., New York (1997)
Benjamin, W.: Das Kunstwerk im Zeitalter seiner technischen Reproduzierbarkeit. In: Zeitschrift für Sozialforschung. Institut für Sozialforschung, Frankfurt (1936)
Bensaude-Vincent, B.: Two cultures of nanotechnology? Int J Philos Chem **10**(2), 68 (2004)
Bensaude-Vincent, B., Newman, W.R.: The Artificial and the Natural: an Evolving Polarity. MIT Press, Cambridge (2007)
Berger, T.W.: In: Mankin, E. (ed.) Plugging into the Brain. University of Southern California Chronicle, Los Angeles (1995)
Bertasio, D.: Studi di sociologia dell'arte. Franco Angeli, Milano (1996)
Bio-Brite. http://www.biobrite.com/windowlite.phtml (2000)
Bio-Vascular, Movements on-line. http://www.ioi.com/index.html
Bischoff, R., Graefe, V., Wershofen, K.P.: Combining object-oriented vision and behavior-based robot control, Report, Institute of Measurement Science, Federal Armed Forces University Munich (1996)
Boas, F.: Primitive Art. Instituttet for sammenlignende kulturforskning, Oslo (1927)

M. Negrotti, *The Reality of the Artificial*, Studies in Applied Philosophy, Epistemology and Rational Ethics 4, DOI: 10.1007/978-3-642-29679-6,
© Springer-Verlag Berlin Heidelberg 2012

Bredekamp, H.: Antikensehnsucht und Maschinenglauben (Nostalgia dell'antico e fascino della macchina, Il Saggiatore, Milano), p. 112. Klaus Wagenbach, Berlin (1996)
Brown, R.P.: Rubber product failure. In: Rapra Review Reports, Shawbury, vol. 13(3), Report 147, p. 8 (2002)
Brown, C.R., Brown, M.B.: Coloniality in the Cliff Swallow, p. 57. The University of Chicago Press, Chicago (1996)
Butler, S.: Erewhon, p. 250. Adelphi, Milano (1988)
Ceserani, G.P.: I falsi Adami, p. 88. Feltrinelli, Milano (1969)
Chang, T.L.: Interview by André Picard for McGill News, Alumni Quarterly. http://www.mcgill.ca/alumni/news/w96/chang.htm (1996)
Churchman, C.W.: The artificiality of science. Review of Herbert A. Simon's book 'The Sciences of the Artificial', Contemporary Psychology, vol. 15(6), June, pp. 385–386 (1979)
Collins, H.M.: Artificial Experts: Social Knowledge and Intelligent Machines. MIT Press, Cambridge (1990)
Committee to Review the National Nanotechnology Initiative. National Materials Advisory Board (NMAB), National Research Council (NRC), 2006. A Matter of Size. Triennial Review of the National Nanotechnology Initiative. The National Acadamies Press, Washington, DC. http://www.nap.edu/catalog/11752.html#toc (2006)
Cooke, J.E.: http://www.pharmacy.ualberta.ca/afpc/afpcproceedingsdocument2000.pdf (2000)
Cordeschi, R.: The discovery of the artificial: some protocybernetic developments 1930–1940. Artif Intell Soc **5**(3), 218–238 (1991)
Cordeschi, R.: La scoperta dell'artificiale, p. 105. Dunod, Milano (1998)
Crandall, B., Lewis, B. (eds.) Nanotechnology Research Perspectives, pp. vii–viii. MIT Press, Cambridge (1997)
Dawkins, R.: Climbing Mount Improbable (Alla conquista del monte improbabile), p. 279. W.W. Norton & Company, Mondadori, Milano (1996)
de Solla Price, D.: Automata and the origin of mechanism. Technol Culture, **1**, 8 (1964)
Denis, M.: Image and Cognition. Presses Universitaires de France, Paris (1989)
Dennett, D.: Cognitive science as reverse engineering several meanings of 'top–down' and 'bottom–up'. In: Prawitz, D., Skyrms, B., Westerståhl, D. (eds.) Proceedings of the 9th International Congress of Logic, Methodology and Philosophy of Science, pp. 679–689. North-Holland Elsevier Science, Amsterdam (1994)
DIP-INFM, University of Genoa. http://www.sts.tu-harburg.de/projects/Esprit-BR/Synopses/Projects/6961.html
Drexler, K.E.: Engines of Creations: Challenges and Choices of the Last Technological Revolution, p. 14. Doubleday, New York (1986)
Fitzgibbons, S.J.: Making artificial organs work. Technol Rev, **97**, 34 (1994)
Edwards, S.A.: Engineering Human Tissue. http://members.aol.com/salaned/writings/engineer.htm (2001)
Flavor Concepts, Inc. http://flavorconcepts.com/bakery.htm
Fryer, D.M., Marshall, J.C.: The motives of Jacques de Vaucanson. Technol Culture **20**, 257–269 (1979)
Fubini, E.: Estetica della musica. Il Mulino, Bologna (1995)
Galletti, P.M., Colton, C.K.: Artificial lungs and blood-gas exchange devices. In: Bronzino, J.D. (ed.) Tissue Engineering and Artificial Organs, pp. 63–66. Taylor & Francis, Boca Raton (2006)
Galloni, M.: Microscopi e microscopie, dalle origini al XIX secolo, Quaderni di storia della tecnologia, 3, Levrotto & Bella, p. 23 (1993)
Gausman, H.W., Ramser, J.H., Dungan, G.H., Earle, F.R., MacMasters, M.M., Hall, H.H., Baird, P.D.: Some effects of artificial drying of corn grain. Plant Physiol, **27**(4), 801 (1952)
Gibbs, W.W.: Artificial muscle. http://cape.uwaterloo.ca/che100projects/organs/Muscles/muscle.htm (1998)
Goldman, C.: Artificial lakes face real environmental conflict. Excerpted from the Davis Enterprise Wednesday, Oct. 6 (1999)

Guy, C., Lesure, F.: Debussy e il simbolismo, pp. 25–40. Palombi, Roma (1984)
Hamdy, R.C.: The thyroid gland: a brief historical perspective. South Med J (2002)
Harnad, S.: Levels of functional equivalence in reverse bioengineering: the Darwinian Turing test for artificial life. Artif Life, **1**(3) (1994)
Haykin, S.: Neural Networks: a Comprehensive Foundation, p. 2. Macmillan, New York (1994)
Hitachi Instruments, Inc., Hitachi Diffraction Gratings. http://www.hii.hitachi.com/grate.htm
Hoffman, W.: Forging New Bonds, Inventing Tomorrow. University of Minnesota, Institute of Technology, Spring, St. Paul (1995)
Hoffmann, R., Leibowitz Schmidt, S.: Old Wine, New Flasks, pp. 19–20. Freeman and Company, New York (1997)
Jefferson, T.: Thomas Jefferson to John Adams, October 28. http://128.143.230.66/members/historical/quotations.asp#natural_vs_artificial_aristocracy (1813)
Johnson-Laird, P.N.: Mental Models. Towards a Cognitive Science of Language, pp. 45–46. Cambridge Universiy Press, Cambridge (1983)
Joy, B.: Why the future does not need us. Wired, 8, April (2000)
Kaneko, W.M., Riley, E.P., Ehlers, C.L.: Effects of artificial rearing on electrophysiology and behavior in adult rats. Depression Anxiety, **4**, 257–331 (1998)
Kawanishi, M.: Developing a method to assess the barrier performance of high-level radioactive waste disposal facilities. Abiko Research Laboratory, Nuclear Fuel Cycle Project Department. http://criepi.denken.or.jp/RD/nenpo/1995E/seika95gen6E.html (1995)
Keaveny, T.: Presentation page for the Berkeley Orthopaedic Biomechanics Research. http://biomech2.me.berkeley.edu/prosthesis.html (1996)
Kern, R.J.: Artificial reefs: trash to treasure. National Geographic News. http://news.nationalgeographic.com/news/2001/02/0201_artificialreef.html (2001)
Kim, M.S., Sur, J., Gong, L.: Humans and humanoid social robots in communication contexts. AI Soc **24**(4), 317–325 (2009)
Kivy, P.: The Corded Shell Reflections on Musical Expression. Princeton University Press, Princeton (1980)
Knight, D.B.: Landscapes in Music Space, Place, and Time in the World's Great Music, p. 10. Rowman & Littlefield Publishers, Lanham (2006)
Köhler, J. F.: Historia Scholarum Lipsiensium (1776), quoted In: David, H.T., Mendel, A.: The Bach Reader. Dent, London (1946), seen in Amis, J., Rose, M.: Words about Music, p. 186. Faber and Faber, London (1989)
Kuhn, T.: The Structure of Scientific Revolutions. Chicago University Press, Chicago (1970)
Kumar, S., Marescaux J. (eds.): Telesurgery, p. 144. Springer, Berlin (2008)
Kurzweil, R.: The Age of Spiritual Machines: When Computers Exceed Human Intelligence. Penguin Books, New York (2000)
Langer, S.: Feeling and Form. Routledge & Kegan Paul, New York (1953)
Landow, G.P.: Hypertext: the Convergence of Contemporary Critical Theory and Technology. Johns Hopkins University Press, London (1992)
Luhmann, N., De Giorgi, R.: Teoria della società. Angeli, Milano (2007)
Langreth, R.: Virtual reality head mounted distress. Pop Sci, **245**(5), 49 (1994)
Langton, C.G.: Artificial life. In: Langton, C.G. (ed.) Artificial Life, Santa Fe Institute Studies in the Science of Complexity, p. 32. Addison-Wesley, Redwood City (1989)
Langton, C.G.: Preface. In: Langton, C.G., Taylor, C., Farmer, J.D., Rasmussen, S. (eds.) Artificial Life II, Volume X of SFI Studies in the Sciences of Complexity, pp. xiii–xviii. Addison-Wesley, Redwood City (1992)
Larson Utility Camouflage. http://www.utilitycamo.com/index.html (2001)
Lee, M., Freed, A., Wessel, D.: (1991) Real-time network processing of gestural and acoustic signals. CNMAT, University of California, internal report
Leiser, D., Cellérier, G., Ducret, J.J.: Une étude de la fonction representative. Archive de Psychologie, **XLIV**, 171 (1976)

Leschiutta, S.R., Leschiutta, M.: I primi strumenti di misura elettrici. Quaderni di storia della tecnologia **3**, 58–59 (1993)
Losano, M.: Storie di automi, p. 128. Einaudi, Torino (1990)
Magnani, L.: Semiotic brains and artificial minds. How brains make up material cognitive systems. In: Gudwin, R., Queiroz, J. (eds.) Semiotics and Intelligent Systems Development, p. 3. Idea Group, Inc., Hershey (2007)
Mahowald, M. A., Mead, C.: The silicon retina. Le Scienze-Sci Am, **275** (1991)
Martin, G.R.: The Visual Problems of Nocturnal Migration, pp. 185–197. Bird Migration, Berlin (1990)
Mazlish, B.: The man-machine and artificial intelligence. Stanf Humanit Rev, **4**(2) (1995)
McGowan Center. http://www.upmc.edu/mcgowan/ArtHeart/Project01.htm
Mikos, A.G., Bizios, R., Wu, K.K., Yaszemski, M.J.: Cell transplantation. The Rice Institute of Biosciences and Bioengineering. http://www.bioc.rice.edu/Institute/area6.html (1996)
Minsky, M.: Will robots inherit the earth? Sci Am (1994)
Moline, J.: Virtual Reality for Health Care: A Survey. National Institute of Standards and Technology, Gaithersburg, 20899
Monge-Nájera J., Blanco M.: Plants that live on leaves. http://www.ots.duke.edu/tropibiojnl/TROPIWEB/BOTANICA/EPIFILOS.htm (2001)
Moravec, H.: The robot as liberation from human nature. Transcript of 1989 Hull Memorial Lecture, Interactions 10, pp. 32–42. Worcester Polytechnic Institute, Worcester, December (1989)
Muzumdar A. (ed.): Powered Upper Limb Prostheses, p. 141. Springer, New York (2004)
Mila, M.: L'esperienza musicale e l'estetica, p. 20. Einaudi, Torino (1950)
Monantheuil de, H.: Quaestiones mechanicae (1517)
Morris, C.: The Open Self, p. 130. Prentice-Hall, New York (1948)
National Institutes of Health.: Clinical applications of biomaterials. NIH Consens Statement, **4**(5), 1–19 (1982)
National Research Council: Materials Science and Technology, pp. 20–21. The National Academic Press, Washington (2001)
Needham, J.: Science and Civilization in China, p. 53. Cambridge University Press, Cambridge (1975)
Negroponte, N.: Essere digitali, p. 123. Sperling & Kupfer, Milano (1995)
Negrotti M. (ed.): Understanding the Artificial (Capire l'artificiale, Bollati-Boringhieri, Torino, 1990, 1993). Springer, London (1991)
Negrotti, M.: Naturoidi: Technology Review, Italian edn. XI, p. 5. Rome (1998)
Negrotti, M.: From the artificial to the art, Leonardo, vol. 32(3). Isast- MIT Press, Cambridge (1999)
Negrotti, M.: Naturoids: on the Nature of the Artificial. World Scientific, Singapore (2002)
Noishiki, Y., Miyata, T.: Polyepoxy compound fixation. In: Wnek, G.E., Bowlin, G.L. (eds.) Encyclopedia of Biomaterials and Biomedical Engineering, vol. 3. Informa Healthcare, New York (2008)
Oppenheim, P., Putnam, I.: Unity of science as a working hypothesis. In: Feigl, H., et al. (eds.) Concepts, Theories, and the Mind-Body Problem, pp. 3–36. University of Minnesota Press, Minneapolis (1958)
Pirandello, L.: Arte e scienza, p. 106. Mondadori, Milano (1994)
Poe, E.A.: The landscape garden. In: The Works of the Late Edgar Allan Poe, vol. IV, pp. 336–345. Redfield, New York (1856)
Poincaré, H.: Mathematical creation. In: Ghiselin, B. (ed.) The Creative Process, p. 33. University of California Press, Los Angeles (1952)
Polanyi, M.: The Tacit Dimension. Doubleday, New York (1966)
Posner, R.: What is culture? Toward a semiotic explication of anthropological concepts. In: Koch, W.A. (ed.) The Nature of Culture. Brockmeyer, Bochum (1989)
Prendergast, P.J.: (2001) Bone prostheses and implants. In: Cowin, S.C. (ed.) Bone Mechanics Handbook, p. 35. CRC Press, Danvers

Qvortrup, L.: Sistemi naturali, sociali e artificiali verso una tassonomia dell'artificiale. In: Negrotti, M. (ed.) Artificialia. Clueb, Bologna (1995)
Rixen, C., Stoeckli, V., Ammann, W.: Does artificial snow production affect soil and vegetation of ski pistes? Rev Perspectives Plant Ecology Evol Syst **5**(4), 219 (2003)
Regge, T.: Infinito, viaggio ai limiti dell'universo, p. 142. Mondadori, Milano (1994)
Rizzotti, M.: Il concetto di artificiale, Memorie dell'Istituto Veneto di Scienze, p. 34. Lettere ed Arti, Venezia (1984)
Rosen, R.: Bionics revisited. In: Karlqvist, A., Svedin, U., Haken, H. (eds.) The Machine as Metaphor and Tool. Springer, New York (1993)
Robinson, J.: Music as a representational art. In: Alperson, P. (ed.) What is Music? An Introduction to the Philopohy of Music. Pennsylvania State University Press, University Park (1994)
Robinson, J.: The expression and arousal of emotion in music. J Aesthet Art Crit, **52**, 13–22 (Temple University, Philadelphia) (1994)
San Jose Mercury-News, Consumer Corner. http://www.gorp.com/gorp/food/lunch.htm (1997)
Schneider, M.: Le rôle de la musique dans la mythologie et les rites des civilisations non europénnes (La musica primitiva, Adelphi Edizioni, Milano, 1992). éditions Gallimard, Paris, p. 35
Sanes, K.: http://www.transparencynow.com/zoos2.htm (1998)
Searle, J.: Minds, brains and programs. Behavioral Brain Sci **3**(3), 417–457 (1980)
Seidensticker, B.: Future Hype: the Myths of Technology Change, pp. 102–103. Berrett-Koehler, San Francisco (2006)
Simon, H.A.: The Sciences of the Artificial (Le scienze dell'artificiale, ISEDI, Milano, 1970). MIT Press, Cambridge (1969)
Sloboda, J.A.: The Musical Mind. The Cognitive Psychology of Music. Oxford University Press, London (1985)
Sneddon, R.: Medical Technology, p. 10. Evans Brothers, London (2008)
Somso Models. http://www.holtanatomical.com/html/SOMSO/ANATOMY-Art_Bone_Models-2.htm
Sternberg, R.J., Lubart, T.: An investment approach to creativity. In: Smith, S.M., Ward, T.B., Finke, R.A. (eds.) The Creative Cognition Approach. MIT Press, Cambridge (1995)
Stravinsky, I.: An Autobiography, p. 53. Marion Boyars, London (1934)
Stravinsky, op cit., p. 54
Suska, S.: Adjusting to nature in cape hatteras national seashore, condensed version of Dolan, R., Hayden, R.: Adjusting to nature in our national seashores, Originally appeared in National Parks & Conservation Magazine, June 1974. http://www.nps.gov/caha/itp/caha_adjust.htm (1998)
Tabor, M.: IGERT statement. http://w3.arizona.edu/~bmpi/profiles/tabor.shtml (1999)
Takimoto Y.: The experimental replacement of a cervical esophageal segment with an artificial prosthesis with the use of collagen matrix and a silicone stent, 6th internet world congress for biomedical sciences, Presentation No. 177 (2000)
Tarde, G.: Les lois de l'imitation (Le leggi dell'imitazione, UTET, Torino, 1976). Alcan, Paris (1890)
Teijin Public Relations & Investor Relations Office. Teijin develops greener process for making superb artificial leather. http://www.teijin.co.jp/english/news/2000/ebd00705.htm (2000)
Texas Water Resources. Replicating Mother Nat, **18**(1) (1992)
Triantafyllou, M.S., Triantafyllou, G.S.: Un robot che simula il nuoto dei pesci. Le Scienze-Sci Am, **3**, 321 (1995)
Tucker, S.A., Nebeker, T.E., Warriner, M.D., Jones, W.D., Beatty, T.K.: Effects of artificial defoliation on the growth of cottonwood simulation of cottonwood leaf beetle defoliation. Gen. Tech. Rep. SRS–71, 169–171 (2004) (Asheville, NC, US Department of Agriculture, Forest Service, Southern Research Station, abstract)
Vallas, L.: Claude Debussy: His Life and Works, p. 172. Dover, New York (1973)

Verheijen, F.J.: The mechanisms of the trapping effect of artificial light sources upon animals. Neth J Zool **13**, 1–107 (1958)

Videm, V.: Neuthrophil-biomaterial interactions. In: Wnek, G.E., Bowlin, G.L. (eds.) Encyclopedia of Biomaterials and Biomedical Engineering, vol. 3. Informa Healthcare, New York (2008)

Vogel, S.: Cats' Paws and Catapults: Mechanical Worlds of Nature and People. W.W. Norton & Company, New York (2000)

von Bertalanffy, L.: General System Theory. Development, Applications. George Braziller, New York (1968)

Weintraub, I.: Pressure used to create artificial diamonds. In: Elert, G. (ed.) The Physics Factbook. http://hypertextbook.com/facts/IleneWeintraub.shtml (1998)

Wiener, N.: Invention: The Care and Feeding of Ideas. MIT Press, Cambridge (1993)

Winslow, R.M.: Ask the expert. How do scientists make artificial blood? How effective is it compared with the real thing? Sci Am (2001)

Wittenburg, C., et al.: Abstract, Universität Hamburg, Institut für Anorganische und Angewandte Chemie. http://www.chemie.uni-hamburg.de/ac/AKs/Dannecker/enviart/html/descript.html (2000)

Wholesale Nutrition. http://www.nutri.com/wn/wn-pl.html (1996)

Woolley, B.: Virtual Worlds. A Journey in Hype and Hyperreality. Blackwell, Oxford (1992)

Young, J.: A Model of the Brain (Un modello del cervello, Einaudi, Torino, 1974), p. 278. Oxford University Press, Oxford (1964)